Big Data Management

The big data paradigm presents a number of challenges for university curricula on big data or data science related topics. On the one hand, new research, tools and technologies are currently being developed to harness the increasingly large quantities of data being generated within our society. On the other, big data curricula at universities are still based on the computer science knowledge systems established in the 1960s and 70s. The gap between the theories and applications is becoming larger, as a result of which current education programs cannot meet the industry's demands for big data talents.

This series aims to refresh and complement the theory and knowledge framework for data management and analytics, reflect the latest research and applications in big data, and highlight key computational tools and techniques currently in development. Its goal is to publish a broad range of textbooks, research monographs, and edited volumes that will:

- Present a systematic and comprehensive knowledge structure for big data and data science research and education
- Supply lectures on big data and data science education with timely and practical reference materials to be used in courses
- Provide introductory and advanced instructional and reference material for students and professionals in computational science and big data
- Familiarize researchers with the latest discoveries and resources they need to advance the field
- Offer assistance to interdisciplinary researchers and practitioners seeking to learn more about big data

The scope of the series includes, but is not limited to, titles in the areas of database management, data mining, data analytics, search engines, data integration, NLP, knowledge graphs, information retrieval, social networks, etc. Other relevant topics will also be considered.

More information about this series at https://link.springer.com/bookseries/15869

Jiawei Jiang • Bin Cui • Ce Zhang

Distributed Machine Learning and Gradient Optimization

Jiawei Jiang
ETH Zurich
Zürich, Switzerland

Bin Cui
School of Electronics Engineering
and Computer Science
Peking University
Beijing, China

Ce Zhang
Department of Computer Science
ETH Zurich
Zurich, Switzerland

ISSN 2522-0179 ISSN 2522-0187 (electronic)
Big Data Management
ISBN 978-981-16-3422-2 ISBN 978-981-16-3420-8 (eBook)
https://doi.org/10.1007/978-981-16-3420-8

This Springer imprint is published by the registered company Springer Nature Singapore Pte Ltd.
The registered company address is: 152 Beach Road, #21-01/04 Gateway East, Singapore 189721, Singapore

Preface

In recent years, with the rapid development of technologies in many industrial applications, such as online shopping, social networks, intelligent healthcare, and IOTs, we have witnessed an explosive increase of generated data. To mine buried knowledge from raw data, machine learning technology is a widely used tool and has become the de facto technique for data analytics in both academia and industry, especially for unstructured data that is beyond the understanding of human beings. Training machine learning models using a single machine has been well-studied, either in a sequential way or in a multicore manner. However, in the data-intensive applications listed earlier, many real datasets can be up to hundreds of terabytes or even petabytes, which have far overwhelmed the computation and storage capability of a single physical machine. Traditional stand-alone machine learning training methods have encountered great challenges accordingly. In order to meet the trend of big data and solve the bottleneck of a stand-alone system, many researchers and practitioners resort to training machine learning models over a set of distributed machines, yielding a new research area in academia—distributed machine learning. Specifically, a class of supervised machine learning algorithms is often solved by a series of iterative gradient-based optimization algorithms, such as stochastic gradient descent (SGD), due to their convergence guarantee and ease of parallelization. These machine learning models include linear regression, logistic regression, support vector machine, gradient boosting decision tree, neural networks, and so on. For these machine learning models, the key to training them in a distributed way is to execute the gradient optimization algorithm in parallel rather than the original sequential execution. Although there is a rich literature on gradient optimization algorithms proposed for distributed machine learning, there is a lack of work that presents a comprehensive overview that elaborates the concepts, principles, basic building blocks, methodology, taxonomy, and recent progress.

This book aims to provide a broad overview of gradient optimization in distributed machine learning, from a systematic perspective, for readers to know the current research topics and start-of-the-art solutions. Through a decomposition of the distributed execution of training machine learning models, we point out the indispensable techniques and then introduce the state-of-the-art methods for each technique individually. After presenting the basics of distributed machine learning, we next present the state-of-the-art distributed gradient algorithms, categorized by the type of machine learning models. These algorithms adopt diverse design spaces regarding the basic techniques to address different challenges in distributed environments. We then introduce the existing distributed machine learning systems according to their architecture and targeted machine learning models.

This book presents the recent developments in a tutorial style, ranging from the indispensable underlying blocks to a range of carefully designed algorithms and systems. Therefore, this book can draw interest from the area of machine learning, artificial intelligence, big data processing, and database management, and it also benefits a broad audience: students in their first years of university, researchers interested in this topic, and engineers who deal with large-scale applications. Undergraduates who are still learning this area can know this research area from scratch, motivating their research interests and strengthening their knowledge foundations. Postgraduates who want to do research in this area can understand the main challenges and find research topics accordingly. Senior researchers in this area can know the state-of-the-art solutions and obtain benefits in their own works. Practitioners can deploy the most recent methods in real applications and solve their problems. Besides, our book is of significant importance for the industry since our targeted topics meet their requirements for the everlasting growth of data generated by real applications.

The required background for understanding this book includes the knowledge of mathematics, programming, and parallel computing, at the level of the first years of university. The knowledge of mathematics includes linear algebra, calculus, and probability theory. The algorithms in this book are generally presented in pseudocode, rather than using a specific language because different languages are adopted in real applications.

Chapter 1 presents the background of distributed machine learning and gradient optimization. Chapter 2 introduces the fundamental blocks in distributed machine learning based on a careful anatomy. Chapter 3 describes the representative gradient optimization algorithms designed for different machine learning models. Chapter 4 gives an overview of the existing distributed machine learning systems that can run gradient optimization algorithms. Chapter 5 concludes this book and offers further resources.

Zürich, Switzerland Jiawei Jiang

Acknowledgments

This book was written based on many works by the DMA group of Peking University and the DS3Lab (Data Sciences, Data Systems, and Data Services) group of ETH Zürich. We would like to thank all the students, engineers, and teachers without whose efforts this book could not have been written.

Zürich, Switzerland Jiawei Jiang

Contents

1 Introduction 1
1.1 Background 1
1.1.1 Methodology of Machine Learning 3
1.1.2 Machine Learning Meets Big Data 4
1.2 Distributed Machine Learning 4
1.3 Gradient Optimization 5
1.3.1 First-Order Gradient Optimization Algorithms 8
1.3.2 Serial Gradient Optimization 11
1.3.3 Distributed Gradient Optimization 11
1.4 Open Problems 12
References 13

2 Basics of Distributed Machine Learning 15
2.1 Anatomy of Distributed Machine Learning 15
2.2 Parallelism 17
2.2.1 Data Parallelism 17
2.2.2 Model Parallelism 20
2.2.3 Hybrid Parallelism 23
2.3 Parameter Sharing 23
2.3.1 Shared-Nothing 24
2.3.2 Shared-Memory 30
2.4 Synchronization 32
2.4.1 Bulk Synchronous Protocol 33
2.4.2 Asynchronous Protocol 34
2.4.3 Stale Synchronous Protocol 35
2.5 Communication Optimization 37
2.5.1 Lower Numerical Precision 37
2.5.2 Communication Compression 43
References 51

3 Distributed Gradient Optimization Algorithms 57
3.1 Linear Models 57
3.1.1 Formalization of Linear Models 59
3.1.2 Overview of Popular Linear Models 60
3.1.3 Single-Node Gradient Optimization 69
3.1.4 Distributed Gradient Optimization 75
3.2 Neural Network Models 83
3.2.1 Formalization of Neural Network 83
3.2.2 Overview of Popular Neural Network Models 86
3.2.3 Distributed Gradient Optimization 91
3.3 Gradient Boosting Decision Tree 101
3.3.1 Formalization of Gradient Boosting Decision Tree 101
3.3.2 Distributed Gradient Optimization 104
References 109

4 Distributed Machine Learning Systems 115
4.1 General Machine Learning Systems 115
4.1.1 MapReduce Systems 115
4.1.2 Parameter Server Systems 123
4.2 Specialized Machine Learning Systems 134
4.3 Deep Learning Systems 138
4.4 Cloud Machine Learning Systems 150
4.4.1 Geo-Distributed Systems 151
4.4.2 Serverless Systems 154
4.5 In-Database Machine Learning Systems 158
References 162

5 Conclusion 167
5.1 Summary of the Book 167
5.2 Further Reading 167
References 168

Acronyms

ASP	Asynchronous protocol
BSP	Bulk synchronous protocol
CNN	Convolutional neural network
DAG	Directed acyclic graph
DL	Deep learning
DRL	Deep reinforcement learning
GAN	Generative adversarial network
GBDT	Gradient boosting decision tree
GD	Gradient decent
LR	Logistic regression
LSTM	Long short-term memory
ML	Machine learning
RNN	Recurrent neural network
SGD	Stochastic gradient decent
SSP	Stale synchronous protocol
SVM	Support vector machine

Chapter 1
Introduction

Abstract Machine learning (ML) has shown its ability to extract buried knowledge from data that is hard for humans to understand. To obtain a satisfying machine learning model, we often use some optimization algorithms to train the model to converged. Among these optimization algorithms, a class of gradient-based optimization algorithms is widely used due to their simple formulations and convergence properties. Since the big data era arrived following the prosperity of data-intensive applications, a new challenge arises when data volume is beyond the capability of one physical machine. To address this scalability problem, distributed machine learning techniques are proposed to train the models with several physical machines. In particular, designing gradient optimization algorithms has become a hot topic in distributed machine learning. In this chapter, we introduce the background of this book, following which we describe the concepts of distributed machine learning and gradient optimization. We finally discuss the challenges and open problems arising recently.

1.1 Background

In the past several decades, the world has experienced an explosive increase in different kinds of data. This is mainly brought by many data-intensive applications, such as social networks, video websites, online shopping, and video surveillance. Figure 1.1 illustrates the data generated within a minute in 2020 by a range of big internet companies [1]. For example, Zoom hosts more than 208,000 participants in meetings, Facebook users upload 147,000 photos, YouTube users upload 500 hours of videos, and Amazon ships 6659 packages. Besides, according to a report of IDC [2], more than 59 zettabytes (ZB) of data will be created, captured, copied, and consumed over the world in 2020. The amount of data created over the next three years will be more than the data created over the past 30 years, and the world will create more than three times the data over the next five years than it did in the previous five years.

This phenomenon is known as the "big data era", which influences people's everyday life in many ways. The term big data does not only indicate the growth

J. Jiang et al., *Distributed Machine Learning and Gradient Optimization*, Big Data Management, https://doi.org/10.1007/978-981-16-3420-8_1

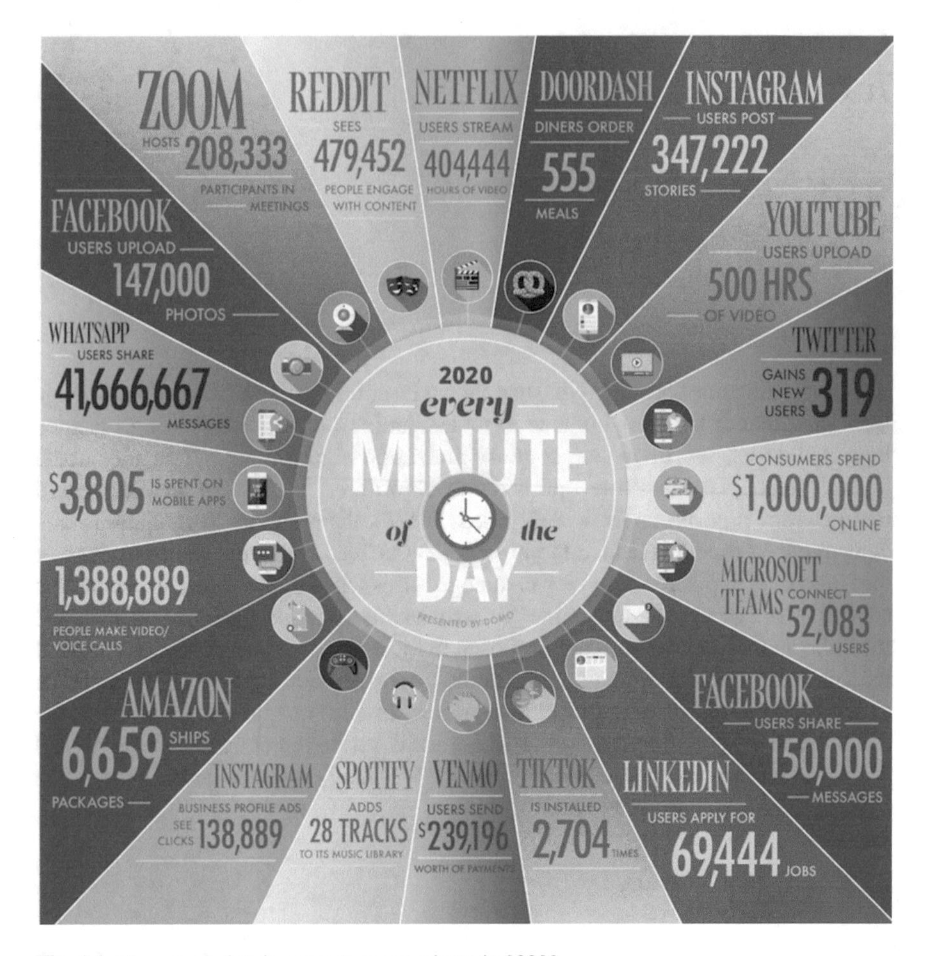

Fig. 1.1 How much data is generate every minute in 2020?

of data but also means it has become the new engine for economic development and social evolution. There are massive potential knowledge and insights mined in big data which we cannot find in small-sized data. One famous story is "beer-diapers"—some supermarkets noticed that customers who bought beers often bought diapers at the same time, so that the managers put them together to increase sales. This reveals strong correlations between real-world objects that can only be found given enough data. The buried knowledge, if explored from big data, can provide enormous commercial value and facilitate everyone's life. As reported by MarketsandMarkets [3], the value of the global big data market will grow from USD 138.9 billion in 2020 to USD 229.4 billion by 2025, at a compound annual growth rate (CAGR) of 10.6% during the forecast period. The potential value of big data has attracted interests from different areas, including academia, industry, and government, to conduct big data analytic.

The analytics of big data can be divided into two categories: simple data analytic and intelligent data analytic. Simple data analytics often adopts OLAP-style (on-line analytical processing) techniques that leverage SQL primitives to handle traditional queries and statistical analysis. However, for large-scale unstructured datasets or users requiring advanced analytic demands, simple data analytic methods are infeasible. In contrast, intelligent data analytic approaches can learn buried information from large-scale data with the help of machine learning methods.

Machine learning is a typical technique that can learn useful knowledge from big data. It is widely used in real-world applications that affect billions of people, such as advertisement recommendation, text mining, multimedia (image, video, speech) processing, financial products, medical treatment, smart cities, and so on. Popular machine learning models contain logistic regression, support vector machine, topic model, decision tree, and neural network. Many research found that [4, 5], the capability and quality of machine learning models improve if given more available data.

1.1.1 Methodology of Machine Learning

Machine learning is a kind of artificial intelligence (AI) technique that has the ability to automatically learn and improve itself from experience without being explicitly programmed. Typically, machine learning builds a statistical model over observations that are already seen, also known as "training data". After the model is generated, the model is used to predict unobserved data in the future. Machine learning models are traditionally categorized into supervised machine learning, unsupervised machine learning, and reinforcement learning, depending on whether the training data is labeled or whether the feedback signal is available.

- *Supervised machine learning.* The training data is labeled, i.e., each input (observation) is given a desired output. Supervised machine learning tries to find a model mapping the input to the output.
- *Unsupervised machine learning.* Unlike supervised machine learning, training data in unsupervised machine learning is unlabeled. Without the guidance of the desired output, the goal is to learn patterns and structures from the input.
- *Reinforcement learning.* For scenarios in which observations are obtained in an interactive manner, reinforcement learning approaches interact with the black-box environment, receive feedback from the environment, and use historical trials to help the next trial.

This book focuses on supervised machine learning since it covers a broad spectrum of machine learning models and is widely used in many real-world applications.

1.1.2 Machine Learning Meets Big Data

The quality of a machine learning model depends on two aspects—the model itself and the training data. The blooming development of distributed machine learning is driven by both aspects—the increase of model complexity and the increase of data volume.

Model Complexity The complexity of a machine learning model fundamentally determines its representational capacity of the model. Especially, many complex models can learn correlations between different features and hence enhance the representational capacity. For example, deep learning models have a powerful ability to approximate data distribution due to their complex model structures including layers with different receptive fields. In contrast, simple machine learning models such as linear models cannot mine correlations between features, resulting in a relatively lower representational capacity. Recently, researchers have contributed a lot in designing more complex models—either wider [6] or deeper [7]. These complex machine learning models bring much more expensive computational cost, e.g., training a deep learning model may cost several days. Although the available hardware (CPU, GPU, etc.) is developing as well, the performance is still not satisfactory for many cases. Distributed machine learning can handle the challenge of training complex models by distributing the computation over several machines.

Data Volume The model quality of supervised machine learning depends on the data since it is built upon it. However, since the size of the training data is limited, it cannot precisely reflect the entire data distribution. Sometimes, the test data and the training data are heterogeneous. In this scenario, the model trained over the training data probably performs inferior over the test data. Even when the training data and the test data are homogeneous, training a complex model over insufficient training data may cause the overfitting problem—the model overfits the training data but cannot generalize to the test data. Fortunately, the rapid growth of available data addresses this problem. Sufficient training data ensure that the data distribution is well represented and the model does not overfit. Since big data have become the fuel of machine learning, the efficient processing of big data is necessary, and distributed machine learning shows its merits accordingly.

1.2 Distributed Machine Learning

Distributed machine learning is a research area that relates to both algorithm design and distributed system design. The algorithm part includes model architecture, training paradigm, and model quality; while the system part needs to resolve common issues in distributed systems, such as parallel computing, network communication, stragglers, and scheduling. The general architecture of distributed machine learning is illustrated in Fig. 1.2. Below we describe the data management and distributed processing individually.

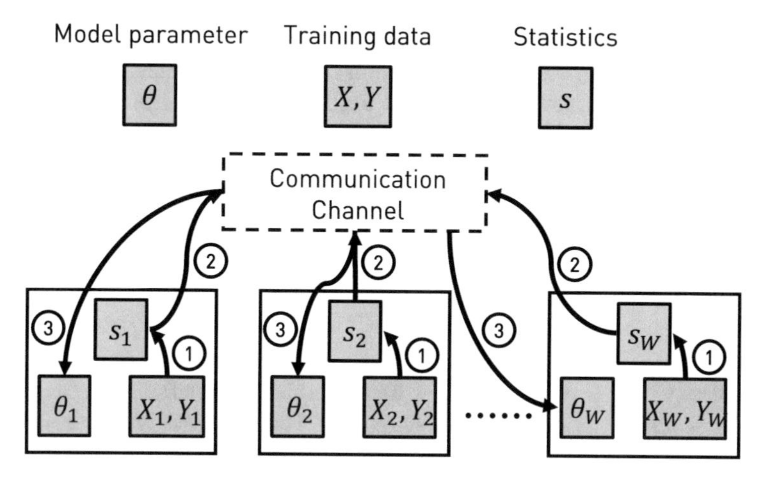

Fig. 1.2 General framework of distributed machine learning

Data Management We use w to denote the model parameter and (X, Y) to denote the training data, where $X = \{x_i\}_{i=1}^N$ is the set of training instances (features) and $Y = \{y_i\}_{i=1}^N$ is the set of desired outputs (labels). There are W machines (also called workers in this book) in this training cluster. Each worker is allocated a subset of the model parameter $\theta_i \subseteq \theta$ and a subset of the training data $\{X_i, Y_i\} \subseteq \{X, Y\}$.

Distributed Processing The training procedure for a machine learning model is as follows: (1) each worker performs calculations over the local model parameter and data, producing some statistics such as gradients; (2) the statistics are aggregated through some communication channel; (3) the local model parameters are updated using aggregated statistics.

1.3 Gradient Optimization

As we focus on a class of supervised machine learning models in this book, we first define the general form of a supervised ML model as the following equation where x_i denotes a training instance, y_i denotes the label of the training instance, and θ denotes the model parameter (often as a vector). $f(x, y, \theta)$ is the objective function, including the loss $L(\{x_i, y_i\}_{i=1}^N, \theta)$ and the regularization term $\Omega(\theta)$. The goal of a supervised ML problem is to find a value of θ that minimizes the objective function f.

$$arg \min_{\theta} f(\theta) = \sum_{i=1}^{N} \underbrace{L(\overbrace{x_i, y_i}^{training\ data}\ ;\ \overbrace{\theta}^{parameter})}_{loss} + \underbrace{\Omega(\theta)}_{regularization} \tag{1.1}$$

The method used to obtain the optimal θ is called the optimization algorithm. Among the existing optimization algorithms, a series of gradient-based optimization algorithms are most widely used. Referring to Fig. 1.2, the statistics in gradient-based optimization algorithms are gradients. These algorithms calculate the derivative (gradient) of θ over the objection function f. Consider θ is a s-dimensional vector $\theta = [\theta_1, \theta_2, .., \theta_s]^T$ in the objective function $f(\theta)$, the gradient of $f(\theta)$ is given by the partial derivatives with respect to each variable in θ:

$$\nabla f(\theta) \equiv g(\theta) \equiv \left[\frac{\partial f}{\partial \theta_1}, \frac{\partial f}{\partial \theta_2}, \ldots, \frac{\partial f}{\partial \theta_s}\right]^T \tag{1.2}$$

The gradient vector is perpendicular to the hyperplane of the function f.

The first-order gradient of a function is a vector, while the "second-order" gradient (or second-order derivative) of a function contains s^2 partial derivatives—the derivative over two different dimensions and the derivative over the same dimension:

$$\frac{\partial^2 f}{\partial \theta_i \partial \theta_j}, i \neq j \;\; and \;\; \frac{\partial^2 f}{\partial \theta_i^2} \tag{1.3}$$

The above second-order partial derivatives together define a square symmetric matrix called Hessian matrix.

$$\nabla^2 f(\theta) \equiv H(\theta) \equiv \begin{pmatrix} \frac{\partial^2 f}{\partial \theta_1^2} & \cdots & \frac{\partial^2 f}{\partial \theta_1 \partial \theta_s} \\ \cdots & & \cdots \\ \frac{\partial^2 f}{\partial \theta_s \partial \theta_1} & \cdots & \frac{\partial^2 f}{\partial \theta_s^2} \end{pmatrix} \tag{1.4}$$

After obtaining the gradient, the model parameter is updated towards the opposite direction of gradients which decreases f. For example, the update rule using the first-order gradient is:

$$\theta_{t+1} = \theta_t - \eta f'(\theta_t) \tag{1.5}$$

where t is the iteration indicator, $f'(\theta_t)$ is the gradient of f at θ_t, and η is a hyper-parameter called learning rate.

These gradient-based optimization algorithms generally adopt an iterative manner until the termination condition is reached, e.g., the predefined maximal iteration or minimal threshold of the objective function. The methodology of gradient-based optimization algorithms assures its convergence guarantees for many problems [8, 9].

According to the type of gradient and the usage of gradient [10], gradient-based optimization algorithms can be classified into the following categories:

- *Steepest descent method.* The simplest, yet effective, gradient optimization algorithm is to use the first-order gradient $g(\theta) \equiv f'(\theta) \equiv \nabla f(\theta)$ as the search direction. The steepest descent method is also called the first-order gradient descent method.
- *Conjugate gradient algorithm.* Instead of using the current gradient, the conjugate gradient algorithm also considers historical gradients. The conjugate gradient at iteration t is computed as:

$$p_{t+1} = \alpha_t p_t - g(\theta_t), \quad \alpha = \frac{g(\theta_k)^T g(\theta_k)}{g(\theta_{k-1})^T g(\theta_{k-1})} \tag{1.6}$$

  Then the algorithm updates θ using the same rule as Eq. 1.5.
- *Newton method.* Another class of algorithms uses second-order gradients to improve the convergence, also known as the Newton method. It can be interpreted taking the second-order term of the Taylor series into consideration:

$$f(\theta + \delta) \approx f(\theta) + g(\theta)^T \delta + \frac{1}{2}\delta^T H(\theta)\delta \tag{1.7}$$

  The above is minimized with respect to δ by setting the differentiating of δ to zero:

$$\delta = -H(\theta)^{-1} g(\theta) \tag{1.8}$$

  Using the same update rule in Eq. 1.5, the model θ is updated in the direction of δ. Although second-order gradient optimization assures a faster theoretical convergence rate, it risks a problem when the Hessian matrix is not positive definite and suffers an expensive cost.
- *Quasi-Newton method.* To achieve a trade-off between the first-order (steepest) method and the second-order (Newton) method, the Quasi-Newton method proposes to approximate the Hessian matrix using the previous first-order gradients. Typical Quasi-Newton methods are DFP [11, 12], BFGS [13, 14], L-BFGS [15], and OWL-QN [16].

Despite the rich research works on gradient optimization algorithms, their deployments over large-scale data are not straightforward. When the data volume and the model complexity become increasingly larger, the methods need to store second-order derivatives or historical first-order derivatives, e.g., the Newton Method and the Quasi-Newton method, are inefficient in terms of computation, memory footprint, and communication. In contrast, first-order gradient optimization yields first-order gradient vector efficiently and has a small memory footprint. Therefore, most, if not all, large-scale machine learning tasks choose the first-order optimization algorithms. This book mainly focuses on first-order gradient optimizations in distributed machine learning.

1.3.1 First-Order Gradient Optimization Algorithms

In this section, we describe several well-known first-order gradient optimization algorithms [17]. According to how much data is used every iteration and how to construct the first-order gradients, we summarize them as follows.

1.3.1.1 Batch Gradient Descent

The vanilla first-order gradient descent is also called batch gradient descent. At every iteration, it uses the entire training data and computes a gradient vector accordingly:

$$\theta_{t+1} = \theta_t - \eta f'(\theta_t; (X, Y)) \tag{1.9}$$

where η is a hyper-parameter called learning rate. Batch gradient descent provides a concrete convergence guarantee to reach the global optimum for convex optimization problems and a local optimum for nonconvex problems. Nevertheless, since each iteration needs to scan the whole dataset, batch gradient descent may be slow for large-scale datasets. Worse, it is infeasible for cases where (1) the dataset that cannot fit in the main memory and (2) data arrive in a streaming pattern.

1.3.1.2 Stochastic Gradient Descent

In contrast to batch gradient descent, stochastic gradient descent (SGD) takes one single instance from the training data and computes a gradient vector over this specific instance:

$$\theta_{t+1} = \theta_t - \eta f'(\theta_t; (x_i, y_i)) \tag{1.10}$$

SGD is computationally efficient as only one instance is involved at each iteration. In addition, the nature of SGD makes it appropriate for the online scenario with on-the-fly incoming data. However, compared with the gradient in batch gradient descent, stochastic gradient brings a high variance, yielding unstable convergence in many cases. Typically, the convergence curve of SGD often fluctuates and vibrates. This is unsurprising because batch gradient descent covers the whole data distribution while stochastic gradient descent cannot. Theoretically, SGD provides the same guarantee of converging to the global optimum for a convex problem if the learning rate is decayed as the training proceeds.

1.3.1.3 Minibatch Gradient Descent

Batch gradient descent leads to a slow computation and stochastic gradient descent suffers large gradient variance, a trade-off between these two methods is to use a min-batch of training instances at each iteration.

$$\theta_{t+1} = \theta_t - \eta f'(\theta_t; (x_{i:i+b}, y_{i:i+b})) \tag{1.11}$$

where b is the mini-batch size. The computation of a gradient is fast if b is appropriately chosen, and the variance of the gradient is much smaller than that of SGD, assuring a stable convergence. Due to the above merits, minibatch gradient descent is the de facto choice for the training of machine learning models.

Despite the effectiveness and efficiency of minibatch gradient descent, some variants are proposed to solve problems such as the choice of learning rate and data sparsity. We describe several classical methods in the rest of this section.

Momentum

The hyperplane of the objective function $f(\theta)$ can be very complex when θ is a multivariate vector. In some areas of the hyperplane, there are "valley"-like surfaces, where the slope towards downhill is flat but the slope towards the ridge is steep. The gradient is therefore small at the downhill dimensions but large at the ridge dimensions. This causes slow convergence approaching the optimality and severe fluctuation between the ridges.

Momentum [18] is proposed to tackle this problem by adding historical gradients to the current gradient.

$$\begin{aligned} m_{t+1} &= \gamma m_t + \eta g(\theta_t), \qquad g(\theta_t) = f'(\theta_t; (x_{i:i+b}, y_{i:i+b})) \\ \theta_{t+1} &= \theta_t - m_t \end{aligned} \tag{1.12}$$

Here γ is the momentum term which is recommended to be set to 0.9. At the downhill slope, the gradients remain the same direction across different iterations, so that the momentum term m_t will accumulate as the algorithm iterates. The convergence towards downhill therefore becomes faster. On the other hand, since the direction of gradients at the ridge slope alters frequently, adding the momentum term reduces the gradient change of these dimensions and alleviates unstable convergence accordingly.

Adagrad

Since the gradient vector is calculated over the dimensions of each instance, zero dimensions in an instance yield zero values for the corresponding dimensions in the gradient. For some sparse training data, the distribution of dimensions is heterogeneous—some dimensions occur frequently while others occur infrequently. As a result, in the gradient vector of a traditional gradient descent algorithm, the values for those infrequent dimensions are small, while those for frequent dimensions are large. Since the same learning rate is used for all the dimensions of a gradient vector, it brings slow convergence for less frequent dimensions.

Adagrad [19] is proposed to address this sparsity heterogeneity problem. Instead of using the same learning rate for each gradient dimension, Adagrad chooses a separate learning rate for each dimension. Briefly, Adagrad records the past change of all gradient dimensions and dynamically adjusts the per-dimension learning rate to be inversely proportional to the past change:

$$\theta_{t+1,i} = \theta_{t,i} - \frac{\eta}{\sqrt{G_{t,i} + \varepsilon}} g_{t,i}, \quad G_{t,i} = \sum_{j=0}^{t} ||g_{j,i}||^2 \tag{1.13}$$

Here $G_{t,i}$ is the square sum of a dimension in all the past gradients, and ε is a smoothing term to avoid division by zero. Unlike traditional gradient descent algorithms, Adagrad does not need to manually tune the learning rate, e.g., decaying the learning rate across iterations. Furthermore, Adagrad automatically perceives infrequent dimensions and increase their convergence by giving a large learning rate.

Adam

Adam [20] is a first-order gradient optimization algorithm that combines the tricks of Momentum and Adagrad. Adam keeps a momentum term m which accumulates past gradients and a squared term of past gradients v.

$$\begin{aligned} m_{t+1} &= \beta_1 m_t + (1 - \beta_1) g(\theta_t) \\ v_{t+1} &= \beta_2 v_t + (1 - \beta_2) g^2(\theta_t) \end{aligned} \tag{1.14}$$

m_t is the moving average of past gradients with a decay factor of β_1. v_t is the moving average of squared past gradients with a decay factor of β_2. Normally, β_1 and β_2 are set to a value close to 1, which incurs a cold-start problem at the beginning of training. To further address this issue, a bias-corrected term is introduced:

$$\begin{aligned} \hat{m}_t &= \frac{m_t}{1 - \beta_1^t} \\ \hat{v}_t &= \frac{v_t}{1 - \beta_2^t} \end{aligned} \tag{1.15}$$

Afterwards, $\hat{m}_t$ and $\hat{v}_t$ are used to update the model parameter:

$$\theta_{t+1} = \theta_t - \frac{\eta}{\sqrt{\hat{v}_t} + \varepsilon} \hat{m}_t \tag{1.16}$$

Regarding the hyperparameters in Adam, the suggested values are—$\beta_1 = 0.9$, $\beta_2 = 0.999$, $\varepsilon = 10^{-8}$. Adam has become the most popular first-order gradient optimization algorithm due to its fast performance and hyperparameter free mechanism.

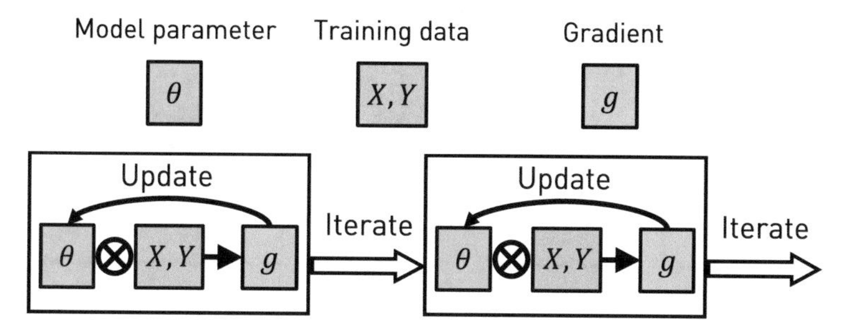

Fig. 1.3 Serial gradient optimization

1.3.2 Serial Gradient Optimization

Before the discussion of distributed gradient optimization, we first show the naive serial implementation of gradient optimization algorithms in Fig. 1.3. The model parameter θ and the training data $\{X, Y\}$ are stored in a single machine. Taking minibatch as an example, the following presents the training process at the t-th iteration:

1. The optimization algorithm takes one minibatch from training data $\{X, Y\}$ at each iteration, with which it calculates the gradient of model parameter θ, denoted by g_t.
2. The model parameter θ is updated by g_t and the learning rate η according to the update rule.
3. The optimization algorithm jumps to step 1 and proceeds to the next iteration.

The serial implementation is quite simple and straightforward. But when a single machine is infeasible or insufficient, researchers and practitioners have to resort to distributed gradient optimization.

1.3.3 Distributed Gradient Optimization

When an ML model solved by a gradient optimization algorithm needs to be handled in a distributed manner, the general framework is illustrated in Fig. 1.4. Assume there are W workers that are physically distributed, each worker is allocated with a subset of the entire training data $X_w, Y_w \subseteq X, Y$, where $w \in \{1, \ldots, W\}$ is the indicator of worker. Likewise, each worker stores a subset of the model parameter $\theta_w \subseteq \theta$. The training process at each iteration is given below:

1. Each worker takes one minibatch from local data X_w, Y_w and computes a gradient g_w using the minibatch and the local model parameter θ_w.
2. The gradients proposed by all the workers are aggregated through the network. The aggregation approach differs for different gradient optimization algorithms.

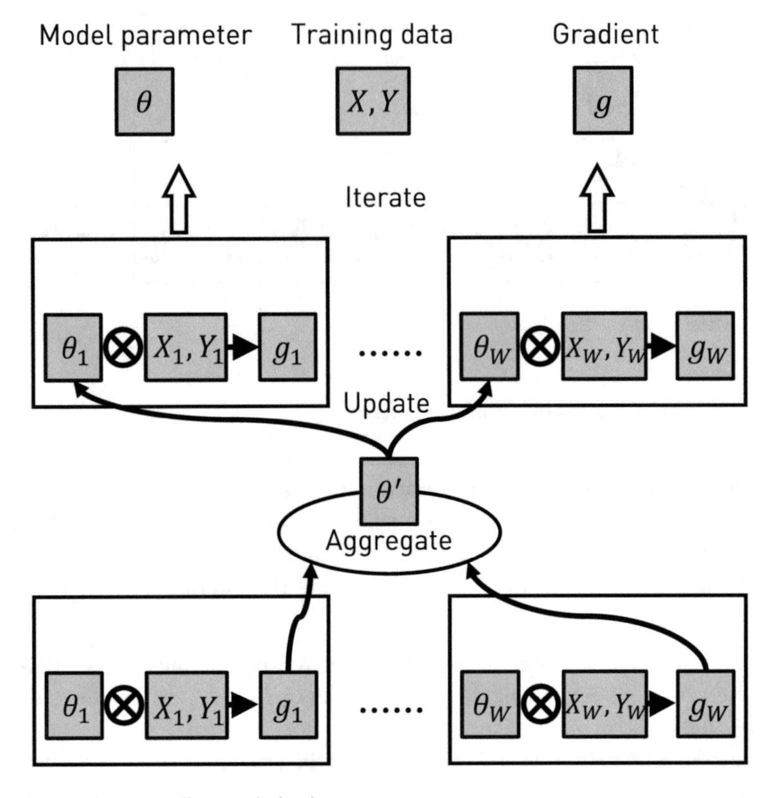

Fig. 1.4 Distributed gradient optimization

3. The local model parameter θ_w is updated with the aggregated gradients.
4. The training proceeds to the next iteration using the new model parameter.

1.4 Open Problems

To design a gradient optimization algorithm in distributed machine learning, there exist several fundamental problems:

1.1 Problem 1: *How can computation be parallelized over distributed machines?*

Since each worker stores a subset of model and training data, the first question is how to appropriately allocate them and perform gradient computation accurately. This needs to consider the capacity of each worker and the computation pattern of the specific algorithm.

1.2 **Problem 2**: *How can model parameters be aggregated from workers?*

A major difference between serial processing and distributed processing is the requirement of aggregating model parameters or statistics (e.g. gradients) from workers. The design of parameter aggregation involves the choice of communication channel, communication framework, and aggregation arithmetic.

1.3 **Problem 3**: *How to synchronize distributed workers?*

A classical problem in distributed data processing is the scheduling of computation units, which is also an important issue for distributed machine learning. Since most gradient optimization algorithms are iterative, a synchronization protocol that schedules the workers is a prerequisite.

1.4 **Problem 4**: *How can communication be optimized during training?*

Distributed machine learning brings data transmission when exchanging gradients. The brought communication overhead could be expensive for many cases, e.g., a large gradient or a slow network bandwidth. For the purpose of accelerating the overall runtime, an efficient solution is to decrease communication. This could be achieved by optimizing the communication framework or reducing the size of transferred data.

References

1. Data Never Sleeps 8.0, https://www.domo.com/learn/data-never-sleeps-8/ (2020)
2. IDC's Global DataSphere Forecast, https://www.idc.com/getdoc.jsp?containerId=prUS46286020 (2020)
3. Big dat market, https://www.marketsandmarkets.com/PressReleases/big-data.asp (2020)
4. Banko, Michele and Brill, Eric: Scaling to very very large corpora for natural language disambiguation. Proceedings of the 39th Annual Meeting of the Association for Computational Linguistics. 26–33 (1995)
5. Brants, Thorsten and Popat, Ashok C and Xu, Peng and Och, Franz J and Dean, Jeffrey: Large language models in machine translation. (2007)
6. Cheng, Heng-Tze and Koc, Levent and Harmsen, Jeremiah and Shaked, Tal and Chandra, Tushar and Aradhye, Hrishi and Anderson, Glen and Corrado, Greg and Chai, Wei and Ispir, Mustafa and others: Wide & deep learning for recommender systems. Proceedings of the 1st Workshop on Deep Learning for Recommender Systems. 5(6), 865–872 (1994)
7. He, Kaiming and Zhang, Xiangyu and Ren, Shaoqing and Sun, Jian: Deep residual learning for image recognition. Proceedings of the IEEE Conference on Computer Vision and Pattern Recognition. 770–778 (2016)
8. Amari, Shunichi: A theory of adaptive pattern classifiers. IEEE Transactions on Electronic Computers. 3, 299–307 (1967)
9. Murata, Noboru and Yoshizawa, Shuji and Amari, Shun-ichi: Network information criterion-determining the number of hidden units for an artificial neural network model. IEEE Transactions on Neural Networks. 5(6), 865–872 (1994)
10. Hicken, Jason and Alonso Juan and Farhat, Charbel: Introduction to Multidisciplinary Design Optimization, http://adl.stanford.edu/aa222/Home.html (2020)

11. Davidon, William C: Variable metric method for minimization. SIAM Journal on Optimization. 1(1), 1–17 (1991)
12. Fletcher, Roger:Practical methods of optimization. John Wiley & Sons (2013)
13. Broyden, Charles George: The convergence of a class of double-rank minimization algorithms 1. general considerations. IMA Journal of Applied Mathematics. 6(1), 76–90 (1970)
14. Liu, Dong C and Nocedal, Jorge: On the limited memory BFGS method for large scale optimization. Mathematical Programming. 45(1-3), 503–528 (1989)
15. Nocedal, Jorge: Updating quasi-Newton matrices with limited storage. Mathematics of Computation. 35(151), 773–782 (1980)
16. Andrew, Galen and Gao, Jianfeng: Scalable training of L 1-regularized log-linear models. Proceedings of the 24th International Conference on Machine Learning. 33–40 (2007)
17. Ruder, Sebastian: An overview of gradient descent optimization algorithms. arXiv preprint arXiv:1609.04747. (2016)
18. Qian, Ning: On the momentum term in gradient descent learning algorithms. Neural Networks. 12(1), 145–151 (1999)
19. Duchi, John and Hazan, Elad and Singer, Yoram: Adaptive subgradient methods for online learning and stochastic optimization. Journal of Machine Learning Research. 12(7) (2011)
20. Kingma, Diederik P and Ba, Jimmy: Adam: A method for stochastic optimization. arXiv preprint arXiv:1412.6980. (2014)

Chapter 2
Basics of Distributed Machine Learning

Abstract Extending the gradient optimization of a machine learning model, which is originally serial, to a distributed setting is nontrivial. Several fundamental techniques are involved in meeting the characteristics of distributed environments. In this chapter, we first conduct an anatomy of distributed machine learning, with which we understand the indispensable building blocks in designing distributed gradient optimization algorithms. Then, we provide an overview for each technique individually and present the state-of-the-art approaches.

2.1 Anatomy of Distributed Machine Learning

Figure 2.1 is the general routine of an iterative gradient optimization algorithm in distributed machine learning. As we have briefly introduced in Chap. 1, there are four fundamental issues yet unsolved—(1) parallelism of computation, (2) aggregation of parameters, (3) synchronization across workers, and (4) optimization of communication. We summarize these four major techniques which are necessary throughout the design of gradient optimization algorithms:

- *Parallelism.* First, we need to parallelize the optimization algorithm by decoupling the optimization problem into several subproblems and deploy each subproblem on one machine (also called worker in this book). Afterwards, the computation is decomposed over all the workers. This needs to allocate (a part of) the model w and (a part of) the training data to each worker. Note that, it is possible to allocate the entire model or the entire training data on one worker.
- *Parameter Aggregation.* Inside one iteration, each worker calculates a gradient vector with local model parameters and training data. Then, the framework aggregates the intermediate parameters (e.g., gradients) and generates a new model parameter before the execution of the next iteration. Note that, the exchanged intermediate parameters can be gradients [1], scaled gradients [2], or the entire model [3]. This requires a parameter aggregation mechanism that aggregates and shares parameters among workers.

J. Jiang et al., *Distributed Machine Learning and Gradient Optimization*, Big Data Management, https://doi.org/10.1007/978-981-16-3420-8_2

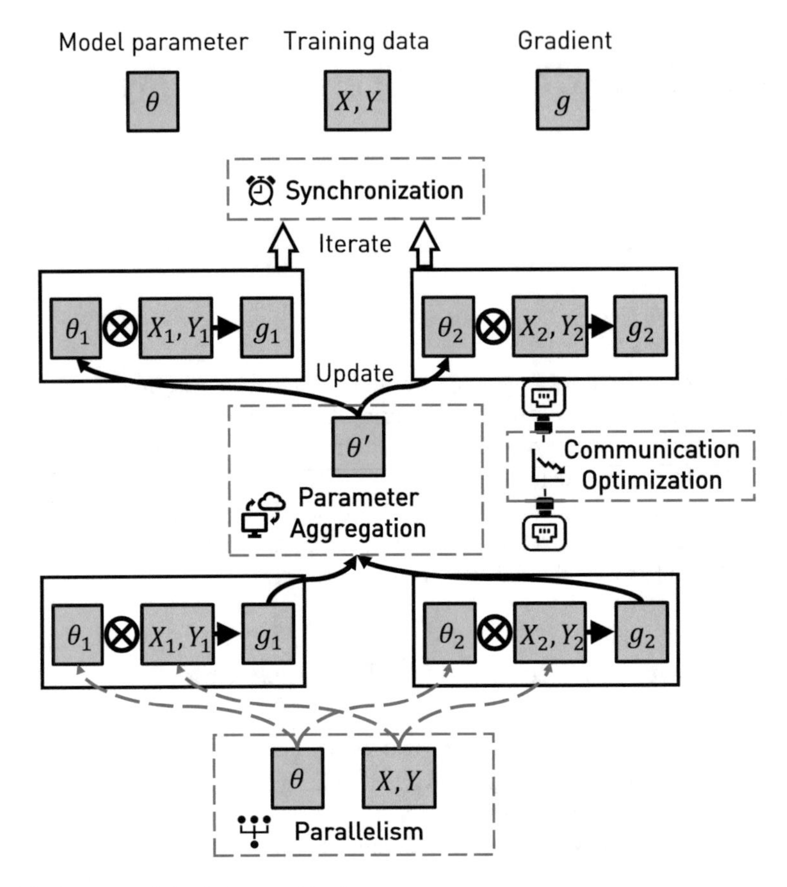

Fig. 2.1 Anatomy of distributed machine learning

- *Synchronization.* For a distributed system, an important problem is scheduling of workers to address computational heterogeneity and communicational heterogeneity. For distributed machine learning, most gradient optimization algorithms are iterative, and therefore synchronization of workers is required across iterations. In particular, in the presence of some stragglers prevalent in distributed environments, the synchronization of stragglers and other workers is important.
- *Communication Optimization.* Compared with serial machine learning, distributed machine learning needs to transfer data through the network, which incurs significant communication costs. To accelerate the overall system performance, an efficient way is to optimize the communication.

2.2 Parallelism

Since distributed machine learning conducts the training over a set of workers, the parallelism of computation is the first design space to be addressed. Specifically, we need to appropriately decouple the optimization problem defined in Eq. 1.1, satisfying two conditions: (1) each subproblem is easier to solve for one worker, and (2) the combination of subproblems is equal to the original optimization problem.

$$arg \min_{\theta} f(x, y, \theta) = \sum_{w=1}^{W} L(\{x_{iw}, y_{iw}\}_{i=1}^{N_w}; \theta_w) + \Omega(\theta) \quad (2.1)$$

Equation 2.1 decomposes the optimization problem of Eq. 1.1 into W subproblems where W is the number of workers. Then, each subproblem is assigned to one worker.

Although the optimization problem is disassembled, the optimization algorithm cannot work unless the training data and model parameters are allocated. The data and model allocation actually depends on how the optimization problem is divided. If one subproblem needs the entire dataset, the training dataset is replicated in every worker; otherwise, the training dataset is partitioned into different partitions, each of which is sent to one worker. Likewise, whether the entire model parameter is partitioned over the workers depends on what each subproblem requires.

The key point of parallelism in distributed ML is how to partition the training data and the model parameter. We assume each subproblem is executed on one worker, given a subset (or the whole set) of the training dataset $\{x_{iw}, y_{iw}\}_{i=1}^{N_w} \subseteq \{X, Y\}$ and a part of (or the whole) model parameter $\theta_w \subseteq \theta$. According to different partition strategies, we summarize the existing parallelism approaches into three categories—*data parallelism*, *model parallelism*, and *hybrid parallelism*.

2.2.1 Data Parallelism

In the big data era, the size of many datasets can be hundreds of GBs, or even up to TB. It is too large to be stored and executed with a single worker; therefore, it is inevitable to launch distributed workers. Figure 2.2 illustrates data parallelism which partitions the training data to different partitions and allocates each partition to the corresponding worker. Meanwhile, all the workers have an entire copy of the model parameter. This strategy is known as *data parallelism*. Data parallelism makes it possible to scale to an extremely large dataset

We assume the training data are a matrix $D = (X, Y)$ by concatenating the training features X and labels Y. Each row in matrix D represents a training instance (x_i, y_i), and each column (except the last label column) represents a feature. According to the partition orientation, there are two different partition approaches—*horizontal partitioning and vertical partitioning*.

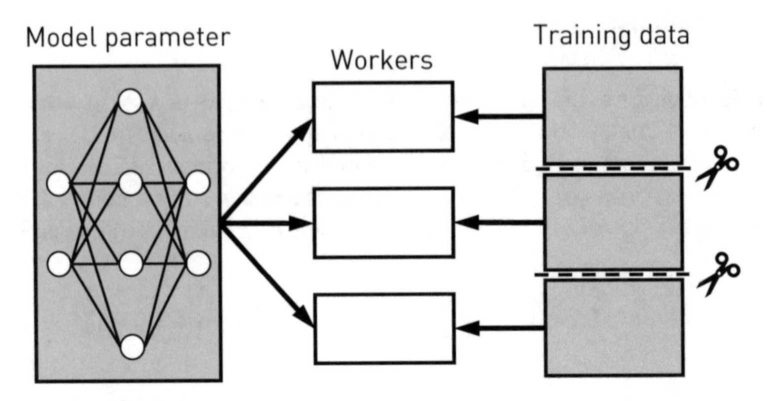

Fig. 2.2 Data parallelism

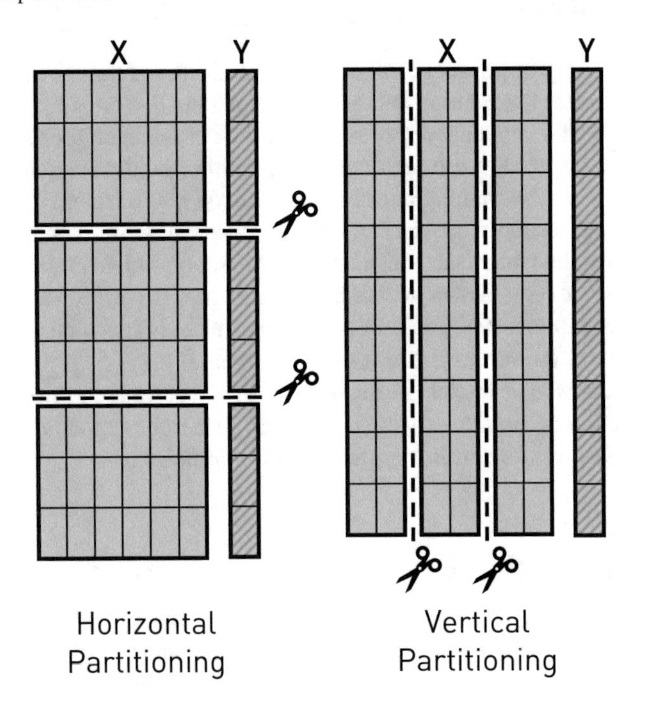

Fig. 2.3 Partitioning strategies in data parallelism

2.2.1.1 Horizontal Partitioning

As the name implies, horizontal partitioning splits the data matrix D horizontally. The left diagram in Fig. 2.3 shows horizontal partitioning. The training data D consists of training features X and desired labels Y. X is represented as a matrix, in which each row is a multidimensional vector, a.k.a. features; while Y is a column vector in which each row is a scalar. Horizontal partitioning strategy divides D

horizontally so that each partition D_w contains a subset of rows in X, denoted by X_w and the corresponding labels in Y, denoted by Y_w. In other words, the training data is partitioned by rows. Concerning the model parameters, each worker stores a model replica with which objective and gradient can be computed.

Recalling the optimization problem in Eq. 2.1, the objective function is as the sum of all the training instances. After horizontal partitioning, each worker calculates the objective over local rows; afterwards, the training system aggregates the objectives of all workers to obtain the final objective value.

With horizontal partitioning, the parallel processing of the gradient optimization algorithm becomes straightforward. Since each gradient is generated using a training instance, each worker computes gradients over its data partition.

$$g_w = f'(\theta; (x_{i:i+b}, y_{i:i+b}) \subseteq (X_w, Y_w)) \tag{2.2}$$

where $(x_{i:i+b}, y_{i:i+b})$ denotes a batch of training instances. The gradients from all the workers are merged through algorithm-specific operators:

$$g = aggregator(g_{1:W}) \tag{2.3}$$

Horizontal partitioning is the most popular strategy in distributed machine learning. PSGD [1] proposes to let each worker perform stochastic gradient descent and launch a master routine to aggregate the solutions from workers. Hogwild! [4] runs SGD in parallel—each processor independently samples training data and concurrently updates a centrally stored model parameter. However, Hogwild! can only run in a single machine and cannot scale to distributed workers. Downpour SGD [5] stores a model replica on each worker, executes minibatch SGD, and pushes the update of the model parameter to a shared in-memory data store called parameter server. XGBoost [6] also partitions training data horizontally, calculates gradients to construct a data structure called gradient histogram, and aggregates gradient histograms. [7] proposes to calculate gradients with a larger batch size over ResNet-50 [8] and ImageNet [9] and uses the *allreduce* collective operator to merge the gradients.

2.2.1.2 Vertical Partitioning

In contrast to horizontal partitioning, another partitioning strategy is vertical partitioning which partitions the training data along with the vertical orientation. The right diagram in Fig. 2.3 presents the partition scheme. As we have defined, each row in matrix X is the features of a training instance and each column represents a specific feature of all the training instances. Vertical partitioning chooses to partition the feature matrix X by columns. With this partitioning strategy, each single training instance (row) is divided and allocated to different workers; and all the items of a specific feature (column) are allocated to a corresponding worker. Regarding the label vector Y, since each label should be companioned with its features, Y is duplicated on each worker without partitioning.

After vertical partitioning, the computation of the objective function becomes more difficult than their horizontal counterparts because a training instance is distributed over different workers. Typically, many machine learning models calculate the dot product between the model parameter θ and the training instance x to obtain the loss $\theta \cdot x$. In the scenario of vertical partitioning, however, each worker is unable to compute the entire dot product itself. Alternatively, assuming one worker w stores columns from wl to wr, each worker computes a partial dot product: $\theta_{wl:wr} \cdot x_{wl:wr}$ which is a scalar. Then, the partial dot products from all the workers are accumulated:

$$\theta \cdot x = \sum_{w=1}^{W} \theta_{wl:wr} \cdot x_{wl:wr} \tag{2.4}$$

The objective of one training instance can be hence obtained using the full dot product.

To compute gradients over the training data, the process is similar to the above. Since the computation of a gradient often needs the dot product of the model parameter and training data, Eq. 2.4 is leveraged to compute the dot product across the workers, with which the gradient is computed according to the specific pattern of the gradient optimization algorithm.

Vertical partitioning is applied in some existing works to accelerate machine learning training. DimmWitted [10] studies training machine learning model with row- and column-partitioned data storage in an NUMA system. ColumnSGD [11] partitions training data by columns, calculates statistics (dot product of model parameter and training instance), sums up the statistics on a master, and performs SGD using merged statistics. Vero [12] studies different partitioning strategies in training GBDT and finds that vertical partitioning outperforms horizontal partitioning on high-dimensional data. For these workloads, the researchers verify the effectiveness of vertical partitioning in decreasing communication.

2.2.2 Model Parallelism

While data parallelism solves the problem of training data being too large by data partitioning, model parallelism is designed for models that are infeasible or inefficient for a single machine. As shown in Fig. 2.4, model parallelism splits the model parameter into several partitions and assigns one partition to the corresponding worker. Meanwhile, each worker can access the whole training dataset. Since one worker only has access to a part of the model parameter, which does not overlap with the other workers, the workers incur no conflicts when they update the model parameter.

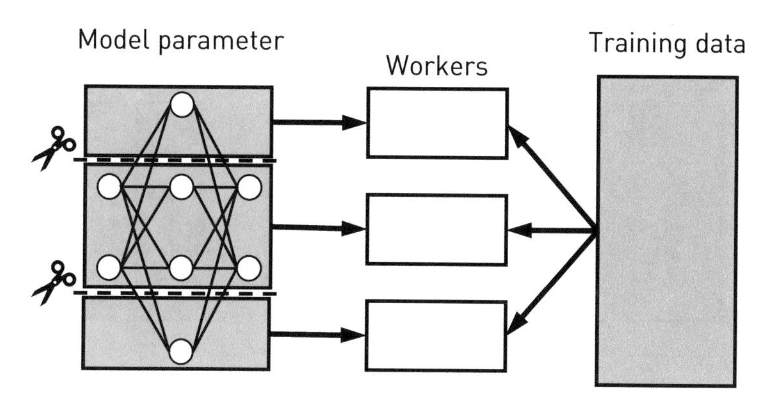

Fig. 2.4 Model parallelism

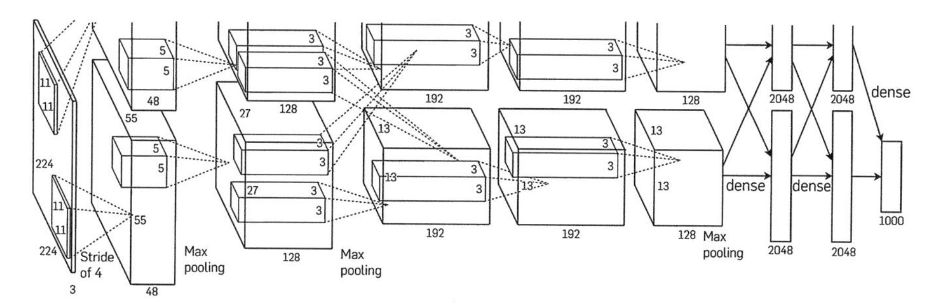

Fig. 2.5 Model parallelism in AlexNet

Model parallelism was widely used in deep learning. Deep learning models are often trained by a graphics processing unit (GPU) due to their highly parallel structure for manipulating computer graphics and image processing. However, the early versions of GPUs only provided limited memory,[1] while the parameter size of a deep learning model could be too large for a single GPU, e.g., AlexNet [13] has more than 60 million parameters, and the full version of the GPT-3 has more than 175 billion parameters [14]. When a single GPU cannot store the entire model parameter, some researchers resort to partition the model parameter over several GPUs. In the implementation of classical convolutional neural network AlexNet [13], the network parameters are partitioned over two GPUs. And each GPU reads data from the entire dataset, computes gradients, and updates the model parameters. Since correlated parameters may be stored in different GPUs, data communication between GPUs often becomes necessary to share their parameters (Fig. 2.5).

In addition to deep learning models, some traditional machine learning models also adopt model parallelism to solve specific problems. Gemulla et al. [15] propose

[1] For instance, the memory of GeForce G100 was 512MB.

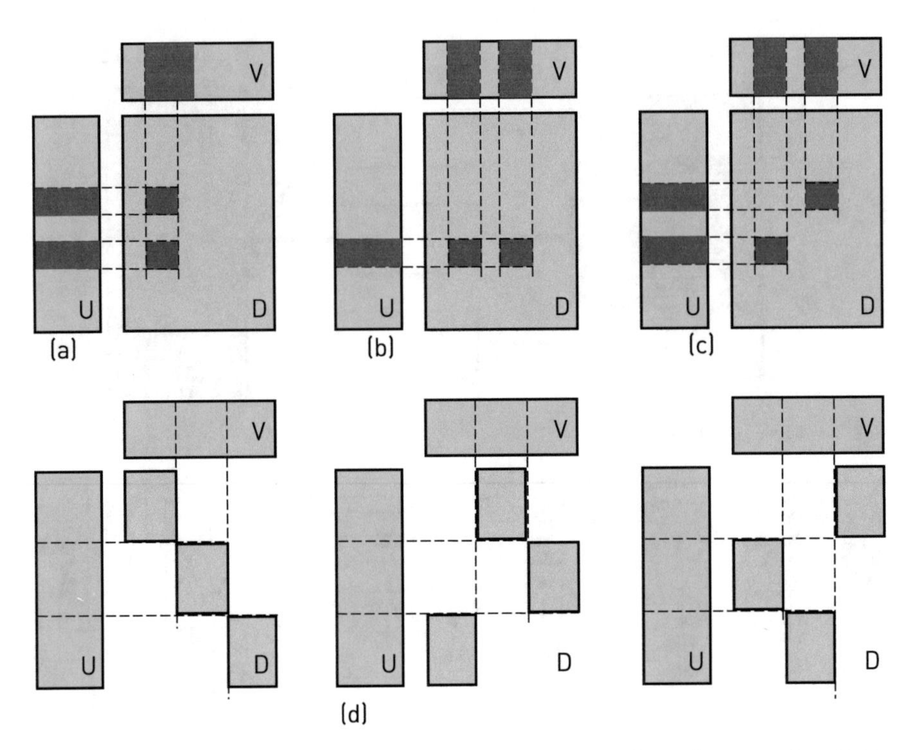

Fig. 2.6 Model parallelism in matrix factorization. (**a**) Column conflict. (**b**) Row conflict. (**c**) No conflict. (**d**) Model parallelism

a model parallelism strategy to train an MF (matrix factorization) model. Matrix factorization algorithm decomposes a high-dimensional user-item matrix D into the product of two lower dimensional matrices—$D \approx U \cdot V$ where U is the user embedding matrix and V is the item embedding matrix. As shown in Fig. 2.6, the training algorithm takes data items from D and updates the corresponding dimensions in U and V. One nature of MF is that the same dimension in the model parameters (U and V) cannot be updated concurrently by two parallel computing units. Therefore, if two data partitions in D have overlapping rows or columns (Fig. 2.6a and b), they may cause conflicts when writing the same parameters. Two computing units can run in parallel only when there is no overlapping row and column (Fig. 2.6c). This work designs a parallel strategy, as shown in Fig. 2.6d, by taking this nature into consideration—(1) in each iteration, the data matrix is partitioned into nonoverlapping partitions; (2) each worker handles a data partition so that the model update has no conflict; (3) proceed to the next iteration using different partitions.

Although model parallelism is not the most popular choice in training ML models, it reveals advantages for some specific cases where the model size is too large or model access is complex. The main disadvantage of model parallelism is that each worker needs to access the whole dataset, which is infeasible for many cases.

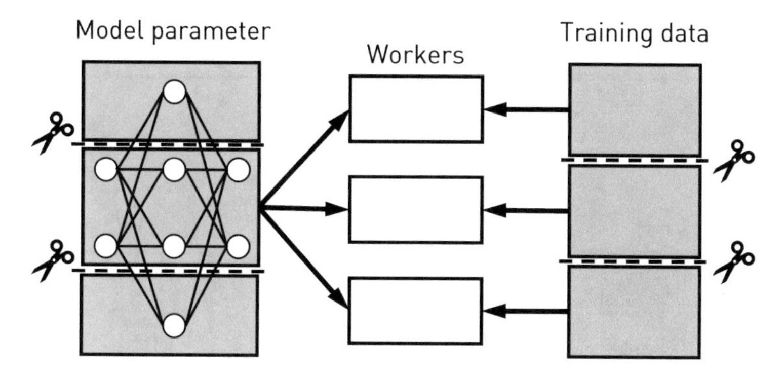

Fig. 2.7 Hybrid parallelism

2.2.3 Hybrid Parallelism

Data parallelism and model parallelism both show their merits for training distributed ML models, one natural intuition is that *we can combine them to obtain benefits from two sides.* Hybrid parallelism is such a strategy that is flexible for different kinds of scenarios. As shown in Fig. 2.7, hybrid parallelism partitions both training data and model parameters. It then assigns a model partition and a data partition to each worker.

SINGA [16] supports hybrid parallelism wherein one worker computes the gradients of a subset of model parameters using a subset of training data, while other workers compute different model parameters and different data partitions. DimBoost [17] designs a hybrid parallelism approach for GBDT (Gradient Boosting Descent Tree) by (1) partitioning the training data over the workers and (2) allocating the tasks of tree splitting to all the workers.

2.3 Parameter Sharing

As shown in Fig. 2.1, a distributed machine learning framework aggregates gradients proposed by all the workers after local computation and updates the local model parameters on workers to a new state. This aggregate-and-update phase needs a model sharing mechanism across the workers to keep the consistency of local model parameters. One way of model sharing is to let each worker maintain a model replica and aggregate gradients through the network, called shared-nothing architecture. In other words, there is not a central node that stores a global state of the model parameters. Alternatively, another way is to maintain a global model parameter in a distributed memory that can be accessed by all workers, called shared-memory architecture.

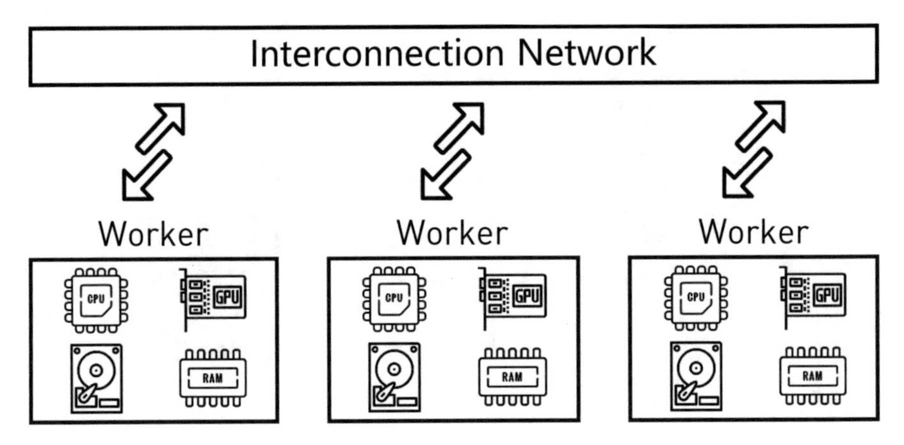

Fig. 2.8 Shared-nothing architecture

2.3.1 Shared-Nothing

A shared-nothing architecture is a distributed architecture in which each computation node has independent processors, memory, and storage (see Fig. 2.8). To keep independent, the nodes do not share their resources with each other. One advantage of shared-nothing architecture is that it can scale-out to a large cluster by simply adding nodes. A typical realization of shared-nothing architecture is sharding in a database, where a table is partitioned to a set of nodes and each node can work independently with the same data schema.

In the context of distributed machine learning, shared-nothing architecture is widely adopted due to its simple abstraction and high scalability. The data management in distributed machine learning naturally satisfies the definition of shared-nothing architecture—the training data and model parameters are partitioned over the workers, and each worker has exclusive hardware resources (processors, memory, disk). Nevertheless, distributed machine learning has its own characteristics—iterative optimization algorithms entail data aggregation among the workers. This can be achieved by some communication infrastructures. Below, we present three popular parallel communication frameworks—MPI (Message Passing Interface), RPC (Remote Procedure Call), and MapReduce. We do not consider some other options, e.g., sockets, because they are seldom used in distributed machine learning, though they have rich histories in the parallel computing community.

2.3.1.1 Message Passing Interface

MPI, short for message passing interface, is a relatively low-level programming API and is widely adopted in parallel computing. MPI makes it possible to pass messages among distributed processes. The first standard of MPI was proposed at Supercomputing'93 conference in 1993 [18, 19], which defined the basic interfaces. Based on this draft of the standard, the first official version of MPI (MPI-1.0) was released in 1994. After that, MPI was widely adopted by industry and academia, due to its high performance, scalability, and portability. After the release of MPI-1.0, the community continuously optimized MPI and proposed new versions. Till now, the latest version of MPI is MPI-4.0.

In the programming model of MPI, an important concept is the `communicator` which defines a group of processes that can communicate with each other. Each process in a communicator is given a rank as a unique identification.

The basic point-to-point interfaces of MPI are *MPI_Send* and *MPI_Recv* operators. A process can send data to another process (identified by rank) via the *send* operator. Likewise, a process can receive data from a specific process via the *recv* operator.

In addition to simple point-to-point operators, many MPI implementations (such as MPICH) provide a series of collective operators for distributed communication [20]. Collective operators involve communication among all processes in a `communicator` group. Typical collective operators include, but are not limited to, `MPI_Bcast`, `MPI_Allreduce`, `MPI_Reduce`, etc. We introduce several classical collective operators below and illustrate them in Fig. 2.9. Each cycle in Fig. 2.9 represents a worker, the number in the cycle represents the rank of the worker, and the rectangle near the cycle represents the data item to be transferred.

- *MPI_Bcast.* The root rank (worker) broadcasts its local data to all other ranks (workers).
- *MPI_Scatter.* The root rank partitions its local data item and sends one portion to each rank independently.
- *MPI_Gather.* The root rank collects all the data items from other ranks, yielding an array of data items.
- *MPI_Reduce.* The root rank collects all the data items from other ranks and sums them together.
- *MPI_Allgather.* Instead of gathering the data items on the root rank, MPI_Allgather gathers the data items on every rank.
- *MPI_Allreduce.* Similarly, all the ranks perform the reduce operation and obtain the sum of all data items.
- *MPI_AllToAll.* MPI_Scatter scatters the data item from the root rank to other ranks, while MPI_AllToAll lets every rank perform the scatter operation.
- *MPI_Barrier.* To schedule different ranks, MPI_Barrier sets a barrier which is a synchronization point for all the ranks.

Researchers and practitioners have designed different MPI implementations following the MPI standard. Open MPI [21] is an open-source MPI implementation that is developed and maintained by a consortium of academic, research, and industry partners. Open MPI combines technologies and resources from several other projects (FT-MPI, LA-MPI, LAM/MPI, and PACX-MPI). MPICH [22] is a high performance and widely portable implementation of the MPI standard. MPICH

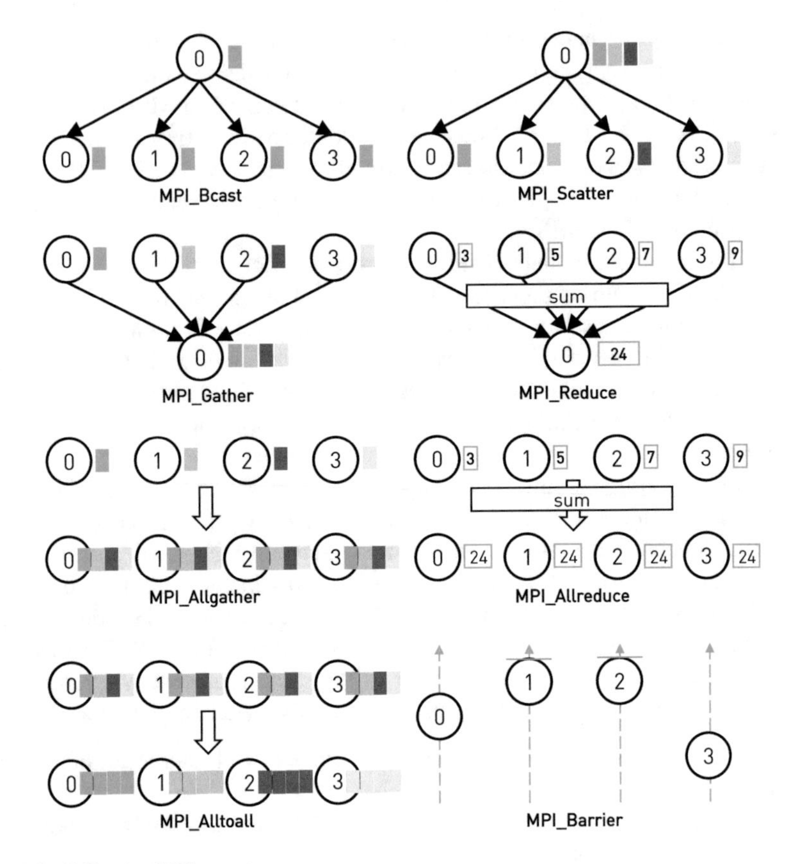

Fig. 2.9 Collective MPI operators

and its derivatives are the most widely used implementations of MPI in the world, especially for supercomputers. mpi4py [23] provides bindings of the MPI standard for the Python programming language, allowing any Python program to exploit multiple processors.

2.3.1.2 Remote Procedure Call

Remote Procedure Call, RPC in short, is a request-response message passing protocol [24, 25]. RPC establishes a server-client framework in which the servers expose their local procedure APIs and the clients can call severs' APIs remotely. The goal of RPC is to let remote server-client communication operate like a local procedure call. RPC facilitates the development of applications due to its multilanguage support, cross-platform deployment, portability, and flexibility. The programming of RPC is user-friendly because it isolates the implementation details relevant to the underlying operating system and networking from the developers.

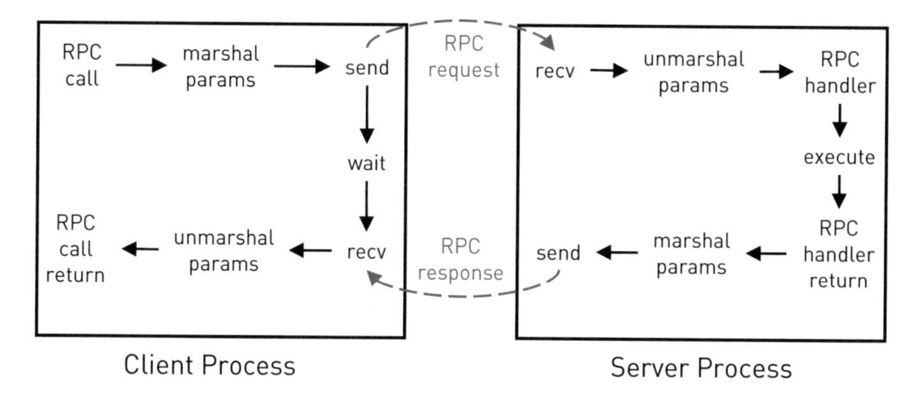

Fig. 2.10 Framework of Remote Procedure Call

Figure 2.10 introduces the lifecycle of one RPC call [26], consisting of the following steps.

1. The client process calls the RPC function and packs desired parameters into a message. This packing operation is called marshaling.
2. The message is sent to the server as an RPC request.
3. The server process receives the RPC request and unmarshals the parameters from the message.
4. RPC handler is triggered to execute the main functionality.
5. The returned result of the RPC handler is marshaled and sent to the client as an RPC response.
6. The client process receives the RPC response, unmarshals the message, and returns the result to the user.

The RPC technique has been studied for decades, resulting in a range of frameworks. gRPC [27] is an open-source high-performance RPC framework that can efficiently connect services in and across data centers with pluggable support for load balancing, tracing, health checking, and authentication. Apache Thrift [28] is a lightweight, language-independent software for point-to-point RPC implementation. Thrift provides clean abstractions and implementations for data transport, data serialization, and application-level processing. Thrift provides an automatic code generation tool that takes a simple definition language as input, generates code across programming languages, and uses the abstracted stack to build interoperable RPC clients and servers. Apache Dubbo is a high-performance Java-based open-source RPC framework. Dubbo [29] offers features such as intelligent load balancing, automatic service registration, high extensibility, runtime traffic routing, and visualized service governance.

2.3.1.3 MapReduce

MPI and RPC, though widely adopted in both industry and academia, are low-level programming APIs. They still require a relatively high programming barrier. The practitioners need to understand the fundamental techniques in designing parallel computing and handling system-related issues, e.g., locality, scheduling, system failures, and stragglers. Under this circumstance, the developers may spend a lot of time in engineering and debugging. The researchers have contributed enormous efforts on reducing the complexities of system design and implementation. Among these works, MapReduce [30] has been acknowledged as the most successful framework in the past decades. MapReduce is a distributed framework for easily writing applications in parallel on large clusters (up to thousands of nodes). MapReduce offers other systematic features such as reliability and fault tolerance.

The programming model of MapReduce is straightforward—the input of the computation is a set of key/value pairs, and the output is another set of key/value pairs. As shown in Fig. 2.11, the key computation abstractions of MapReduce are *map* function and *reduce* function. *Map* function takes an input item and generates a set of intermediate key/value pairs. The framework then groups together related intermediate values with the same intermediate key and sends the grouped results to the *reduce* function. The *reduce* function handles each intermediate key independently—it (1) takes an intermediate key and the values for this key, (2) merges these values to generate another set of values.

The principle of MapReduce easily supports iterative algorithms by executing map and reduce functions sequentially. The adoption of MapReduce for distributed machine learning is also feasible because the involved data, e.g., objective and

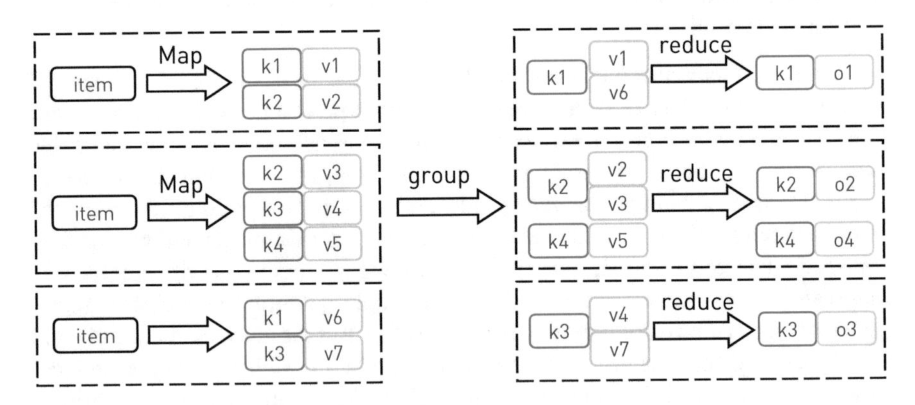

Fig. 2.11 Illustration of MapReduce abstraction

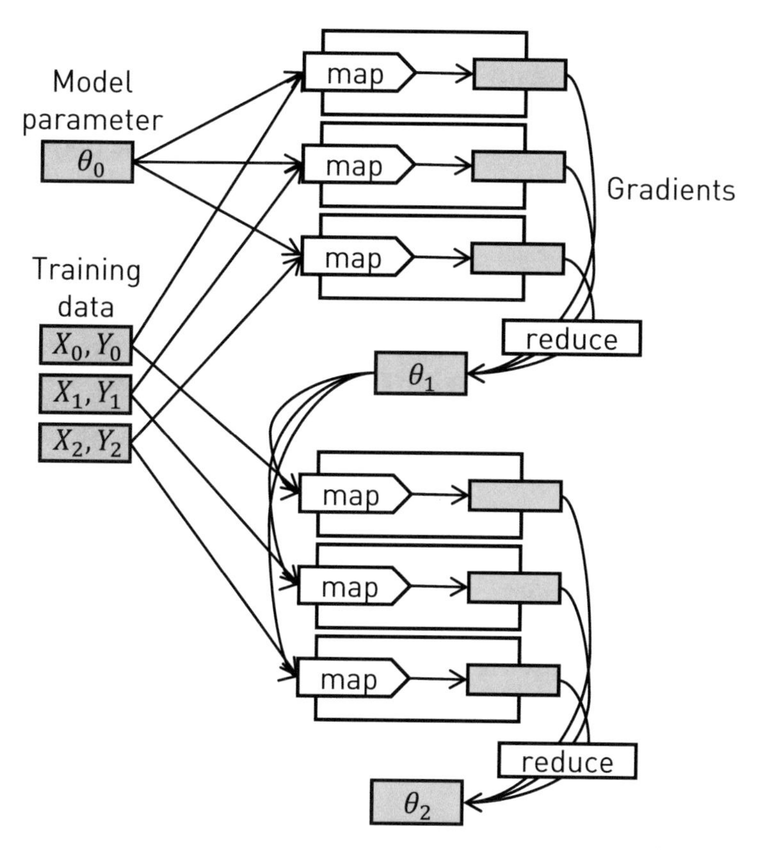

Fig. 2.12 Illustration of distributed machine learning over MapReduce

gradient, are often statistics that can be formed as aggregation:

$$f(\theta) = \sum_{i=1}^{N} L(x_i, y_i; \theta)) + \Omega(\theta))$$

$$g(\theta)) = \frac{1}{N}\sum_{i=1}^{N} f'(x_i, y_i; \theta)) \tag{2.5}$$

The computation of statistics over large-scale data can be parallelized by map function, and the aggregation of statistics can be implemented by reduce function.

Figure 2.12 shows distributed machine learning over the MapReduce framework. The input is the training data and the initial model parameter. The process of iterative training is as follows:

- *Map phase.* The map function on each worker takes model parameter w and its corresponding data partition (X_i, Y_i) as the input. Intermediate statistics such as gradients are calculated with the model parameter and training data.

- *Reduce phase.* The local gradients computed by all the workers are aggregated to obtain the global gradient with which the new model parameter is produced. The next iteration proceeds with the new model parameter.

MapReduce-based machine learning has been extensively explored, yielding a range of solutions [31–33]. Apache Mahout [32] is a project providing distributed and scalable machine learning algorithms focused primarily on linear algebra. Mahout was originally built on Apache Hadoop [34], the de facto choice of MapReduce platform; while it is now migrated to Apache Spark [35], a MapReduce-based platform using Resilient Distributed Dataset [36]. MLlib [37] is the scalable machine learning library of Apache Spark, which covers a broad spectrum of machine learning models, e.g., classification, regression, clustering, decision trees, recommendation, topic model, and ML pipelines. Spark offers multilanguage support and is deeply integrated with many open-source projects (HDFS, HBase, Hive, Yarn, Mesos, Kubernetes).

2.3.2 *Shared-Memory*

In contrast to shared-nothing architecture, the shared-memory architecture in Fig. 2.13 chooses to establish in-memory storage that is accessible for all workers. The workers can read and write shared data in the shared-memory through the interconnection network. Shared-memory has two settings—(1) single machine,

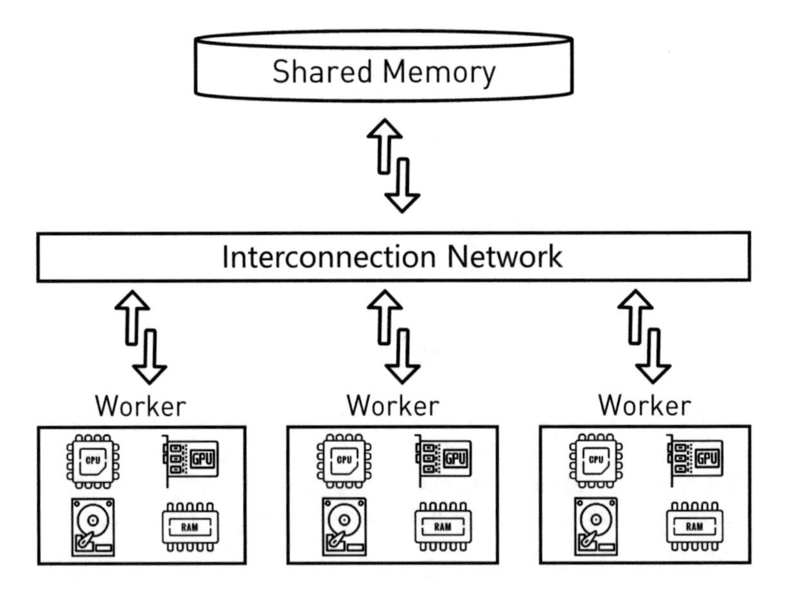

Fig. 2.13 Shared-memory architecture

multiple processes and (2) multiple machines. Regarding parallel machine learning in a shared-memory multicore machine, the community has studied for decades and contributed influential works such as Libsvm [38], Hogwild! [4], Dimmwitted [10], xlearn [39], etc. However, since multicore machine learning is not the topic of our work, we focus on distributed shared-memory architecture for the setting of multiple machines [40].

Shared-memory architecture is proposed for distributed machine learning to solve the problem of shared-nothing architecture. As we have previously discussed, MPI and RPC entail too many complexities of communication details for developers. Although MapReduce lifts the programming efforts from users by designing user-friendly APIs and isolating communication details, it reveals weaknesses facing large-scale training data. In the past, the training dataset was often both small-sized and low-dimensional. However, many training datasets in recent years are large-sized and high-dimensional. Therefore, the generated local statistics on workers are high-dimensional vectors or matrices. When training such datasets with MapReduce, the reduce phase requires one worker to aggregate the local statistics. This aggregation worker may become a bottleneck due to limited network bandwidth. To resolve the single-node bottleneck in a MapReduce-based system, one solution is to assign multiple servers to conduct the aggregation instead of one worker. Specifically, each server creates a shared-memory data store, and the servers together handle the aggregation of statistics. This shared-memory architecture is also called parameter server [5].

Figure 2.14 shows the architecture of the parameter server. The model parameters are partitioned over several machines, also called servers, and hence each server stores a partition of the model parameters. The servers together handle the management of model parameters and provide interfaces for the workers, e.g., *push* and *pull*. Each worker has a local copy of the model parameter and generates

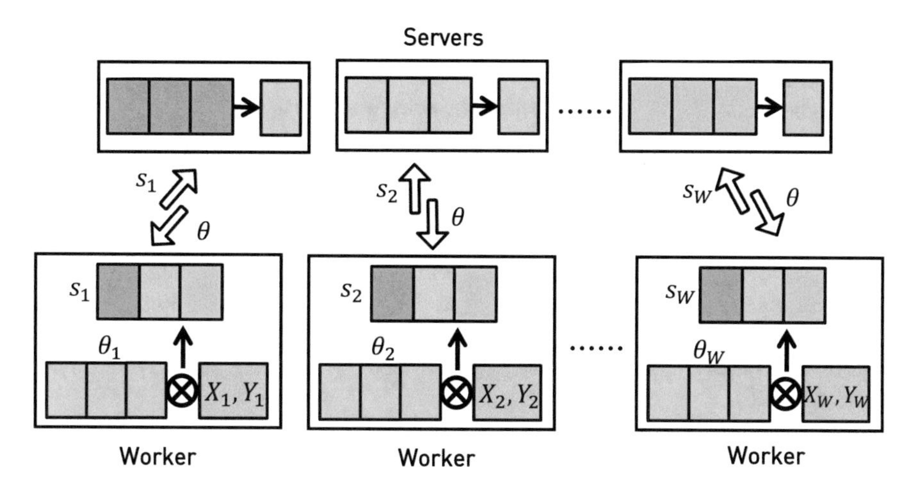

Fig. 2.14 Parameter server architecture

intermediate statistics using the local training data. Then, the same partitioning strategy is performed to the local statistics, and each partition of statistics is sent to the corresponding server through the *push* interface. The server side receives partitions of statistics from workers and updates the partition of model parameters accordingly. When the new state of model parameters is ready on the servers, the workers obtain the new model parameters via the *pull* interface. Note that, the transferred statistics can be gradients, updates of model parameters, or local model parameters [2, 42].

Parameter server prevents a single-point bottleneck because the burden of aggregation is distributed over multiple machines. Parameter server shows superior scalability for large-scale high-dimensional datasets—(1) adding more workers for larger datasets and (2) adding more servers for higher dimensions.

YahooLDA [42] implements a distributed training system for high-dimensional LDA using multiple Memcached servers.[2] Distbelief [5] first proposes the concept of the parameter server and validates its effectiveness with SGD and L-BFGS. Petuum [43] designs a general parameter server platform that supports both data parallelism and model parallelism with a scheduler abstraction. Angel [2] implements a parameter server training engine in the Hadoop ecosystem and provides various machine learning models.

2.4 Synchronization

When training machine learning models in a distributed setting, synchronization protocol is indispensable for scheduling running workers. During the execution of an iterative optimization algorithm, the workers independently generate statistics such as gradients that are aggregated before running the next iteration. Under this training mechanism, the workers should be managed by a synchronization method that assures the local statistics are properly aggregated and the workers are scheduled across iterations.

Another challenge in a real distributed environment is heterogeneity, that is, the computation and communication capabilities of physical machines are different due to different hardware equipment. Even when the machines are equipped with the same hardware, their execution capabilities can still vary as a result of unexpected system interrupt or network congestion.

There is a broad spectrum of synchronization protocols, ranging from pure synchronous to pure asynchronous, according to different tolerance of delay and convergence flexibility. We summarize them into three categories—Bulk Synchronous Protocol (BSP), Asynchronous Protocol (ASP), and Stale Synchronous Protocol (SSP).

---

[2] https://memcached.org/.

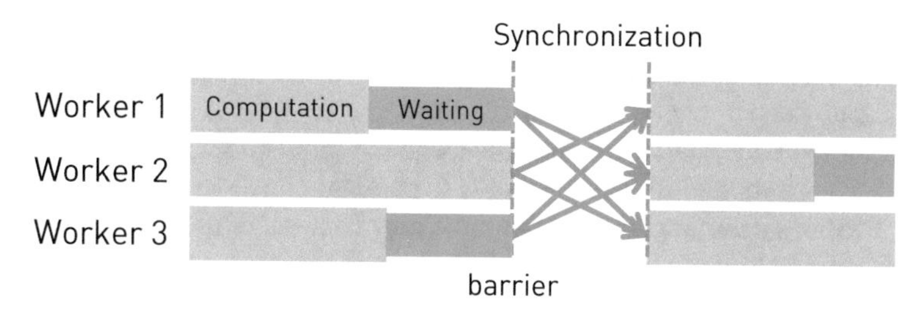

Fig. 2.15 Bulk synchronous protocol

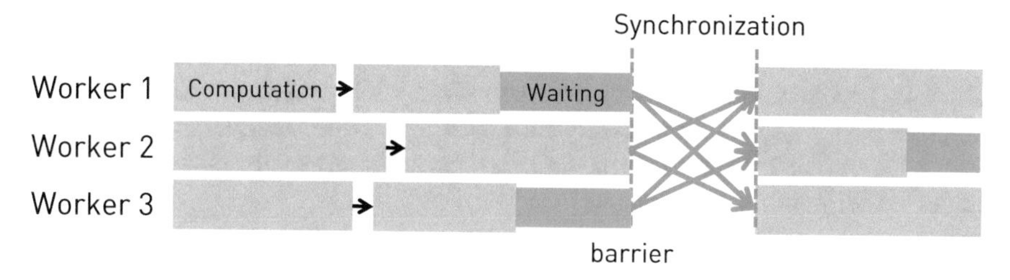

Fig. 2.16 Arbitrarily sized bulk synchronous protocol

2.4.1 Bulk Synchronous Protocol

Figure 2.15 is an example of bulk synchronous protocol (BSP). Under BSP, each worker maintains a copy of model parameters and uses allocated training data partition to compute local gradients. There is a synchronization barrier at the end of each iteration, at which each worker stops running and waits for other workers. The synchronization rule is strict—no worker is allowed to run the $(t + 1)$-th iteration until all the workers finish the t-th iteration. Once the synchronization condition is met, the workers aggregate the local statistics, calculate the model parameters, and continue training with the latest model state. Taking PSGD [1] as an example, each worker executes an iteration of SGD, updates the local model copy; the system aggregates all local model copies and takes their average as the new model parameter.

Cui et al. [44] propose a variant of BSP called A-BSP (arbitrarily sized bulk synchronous protocol). A-BSP sets a rule that the workers can synchronize once every multiple iterations. Figure 2.16 is an example of A-BSP in which the workers perform synchronization every two iterations. A-BSP can save considerable cost on communication as a result of less frequent synchronization.

A special case of A-BSP is the model average (MA) [3, 45]. MA chooses to synchronize every epoch (scan the entire training dataset for one loop). The model average strategy can achieve a trade-off between statistical convergence and system efficiency.

BSP has the following advantages: (1) BSP fits most optimization algorithms for machine learning models, including generalized linear models, clustering, tree models, and deep learning models; (2) the logic of BSP is simple and easy to implement; (3) BSP guarantees equivalence to the sequential execution of the optimization algorithm and therefore assures theoretical convergence. Owing to these merits, BSP has been implemented in most prevalent distributed machine learning systems, such as Mahout [32], MLlib [37], TensorFlow [46], and PyTorch [47].

However, BSP entails a strong dependency among the workers—the execution of one iteration has to wait for the completion of the last iteration. In a heterogeneous distributed environment; however, the execution speed of workers varies due to diverse computation capabilities, network bandwidths, or unpredictable congestion. Hence, workers often need different time cost to run an iteration. This causes a straggler problem, that is, some workers run fast, while others run slow. The faster workers have to spend unnecessary time waiting for the stragglers. As shown in Fig. 2.15, worker 1 first finishes the first iteration and waits for the other two workers at the barrier point.

2.4.2 Asynchronous Protocol

To address the problem of BSP in heterogeneous environments, some researchers resort to Asynchronous Protocol (ASP). Different from BSP, ASP is a best-effort protocol as shown in Fig. 2.17. There is no synchronization barrier at all in ASP so that the workers do not need to wait for each other. Regarding the underlying communication framework, MPI cannot support ASP due to the synchronized nature of collective operators, while the parameter server architecture is compatible with both BSP and ASP. We take the parameter server and minibatch SGD as an example to elaborate ASP: (1) each worker computes local gradients using a mini-batch of training data; (2) each worker partitions the local gradients and pushes them to the parameter server after each iteration finishes computation; (3) each parameter server receives gradients from the workers and updates the stored model parameter according to the rule of SGD; (4) each worker pulls the latest model parameters from the parameter server and starts the next iteration without waiting.

Hogwild [4] lets multiple threads update shared model parameters in a single-node multicore machine. It adopts ASP during the training but cannot be extended

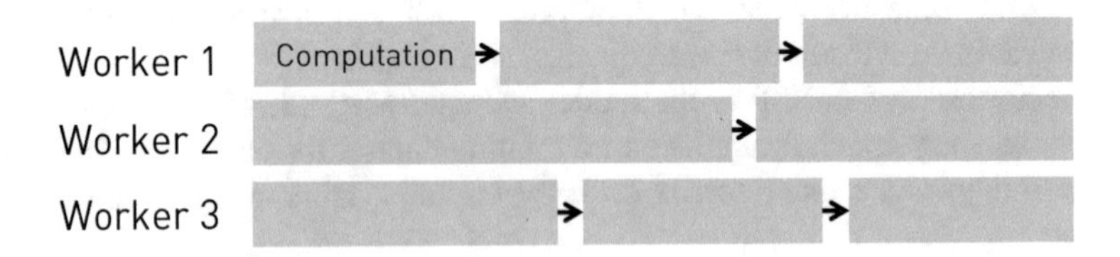

Fig. 2.17 Asynchronous protocol

to a distributed setting. DistBelief [5] implements ASP-based distributed mini-batch SGD over parameter server architecture. Some works [48–50] implement ASP-based SGD in which the workers push local gradients to the parameter server and the servers update the centrally stored model parameters with a scaling factor called the learning rate. DistBelief [5] uses minibatch SGD to update the local model copy and pushes the update of model parameters to the parameter server. This is equivalent to pushing the gradients and multiplying them by the learning rate.

ASP solves the straggler problem in BSP and entirely removes the synchronization barriers. In this manner, faster workers do not waste time on waiting for those slower workers, and the overall system efficiency is accelerated. However, since ASP allows different processing paces across workers and inconsistency between the local model copies, ASP cannot guarantee the correct convergence of gradient optimization algorithms and sometimes even incurs divergence of optimizations.

2.4.3 *Stale Synchronous Protocol*

As discussed above, the performance of BSP is affected by the straggler problem so that the whole system is stalled by the slowest worker. In reality, the straggler phenomenon is very common owing to heterogeneous hardware, hardware failure, data imbalance, resource sharing, and network delay. ASP resolves system heterogeneity by completely removing synchronization barriers but introduces unstable convergence. Therefore, neither BSP nor ASP are ideal synchronization protocols for distributed machine learning.

An alternative solution is to take a trade-off between BSP and ASP—introducing certain flexibility to synchronization other than entirely removing the barriers. This class of protocols restricts the speed gap (e.g., number of iterations, number of epochs, number of used batches) between the workers. In terms of model update, they work in the same way as ASP and push intermediate statistics to the parameter server.

Langford et al. propose a training paradigm in which each worker updates the model parameters in a round-robin fashion, which we call RR-SSP [51]. As illustrated in Fig. 2.18, RR-SSP sorts the workers and sequentially chooses one

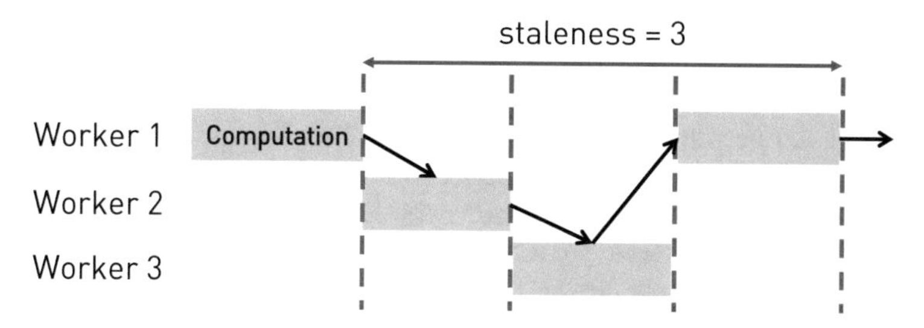

Fig. 2.18 Round-Robin stale synchronous protocol

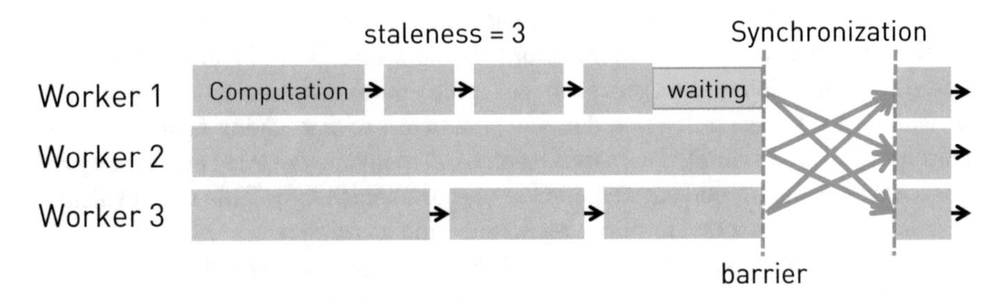

Fig. 2.19 Stale synchronous protocol

worker at a time to update the model parameters with gradients. If there are W workers, one specific worker reads the model parameters every W rounds. That is to say, RR-SSP introduces a constant delay of $\tau = W - 1$ for the workers. This process restricts the inconsistency between the local copy of model parameters and the global model parameters and thereby reduce unstable convergence. However, allowing only one worker runs at a time causes resource idleness, making RR-SSP unfit for a distributed setting.

Ho et al. [52] propose a restricted synchronization protocol for distributed machine learning, called SSP (stale synchronous protocol), as shown in Fig. 2.19. Different from BSP and ASP, the synchronization rule of SSP is that the largest gap of iterations between the fastest worker and slowest worker cannot exceed a predefined threshold s. Consequently, the synchronization barrier is not set at each iteration; instead the barrier is dynamic. The fastest worker needs to wait for the slowest worker when reaching the barrier. Under SSP, the processing on each worker is the same as ASP, and the difference is the implementation of the parameter server:

1. The parameter server monitors the iteration rounds of all the workers.
2. When the parameter server receives statistics from the workers via the *push* function, it directly updates the model parameters on the parameter server using the statistics.
3. When one worker tries to pull the latest model parameters from the parameter server, the parameter server first checks the current iterations of the workers. If the worker is allowed to process the next iteration according to the rule of SSP, the parameter server sends the latest model parameter to the worker. Otherwise, if the querying worker is too fast, the parameter will hang up the pull request and wait until the SSP condition is satisfied.

Dai et al. further introduce an SSP variant called Eager-SSP (ESSP) [53]. In the implementation of SSP, the workers use the *pull* interface of the parameter server to acquire the latest model parameters. ESSP leverages a more aggressive strategy—the parameter server proactively pushes the latest model parameters to the workers once the model parameters on the parameter server are updated. Through this strategy, ESSP decreases the inconsistency and delay between the local model parameters on the workers and the global model parameters on the parameter server. Therefore, ESSP accelerates the convergence of gradient optimization algorithms, at the expense of extra communication costs.

Compared with BSP and ASP, SSP-style approaches can reduce the time consumption waiting for stragglers in BSP and provide theoretical convergence guarantees by bounding the speed gap that ASP fails to guarantee. Although SSP successfully achieves a trade-off between BSP and ASP, it still has inevitable inconsistency between local model parameters and global model parameters. Choosing a small threshold of s can decrease this inconsistency but increase the time spent on waiting. In contrast, choosing a large threshold of s can decrease the overhead of waiting and increase the model inconsistency. Therefore, choosing a proper value of the threshold s becomes a dilemma—it is a case-by-case workload that should be tuned for each training job.

2.5 Communication Optimization

A major difference between single-node machine learning and distributed machine learning is the extra cost of transferring data among workers. As reported by previous works [3, 54, 55], the communication cost sometimes dominates the overall cost. When the computation speed is hard to improve, optimizing the communication is an effective approach to accelerating the execution. This can be achieved by either compressing the transferred data (e.g., gradients) or dropping less important data items.

2.5.1 *Lower Numerical Precision*

Normally, the standard training of machine learning models typically chooses 32-bit (a.k.a `float` or single precision) or 64-bit (a.k.a `double` or double precision) floating-point representation of real numbers for both the model parameters and intermediate statistics. The adoption of more bits preserves the precise representations of the transferred numbers and precise computation over the numbers. Therefore, the high precision representation is suitable for error-vulnerable tasks that require that the outputs are reproducible. Nevertheless, many machine learning models have the nature of being durable to certain errors of model parameters and intermediate statistics during the training process. Furthermore, some research works have shown that introducing noise can improve the performance of machine learning models, especially for nonconvex deep learning models that may benefit from noisy gradients for jumping out of local optimality [56]. Due to the error-resiliency of machine learning models, the researchers resort to using fewer bits to represent the model parameters and/or intermediate statistics for both computation and communication.

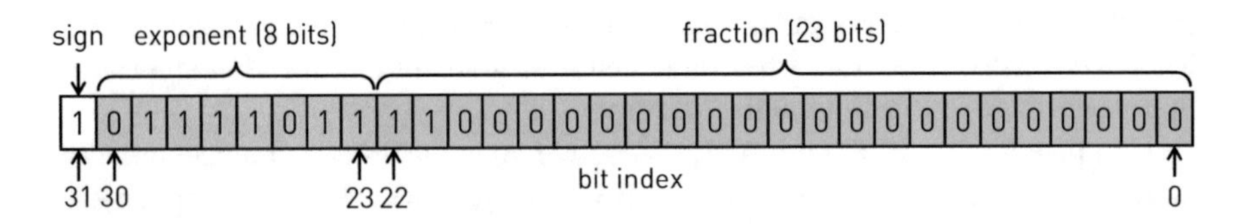

Fig. 2.20 Format of floating-point number

The benefits brought by lower precision are threefold. First, the computation over lower-precision numbers is faster than that of higher-precision numbers. Many advanced hardware and computation frameworks [46, 47] have already provided support for lower-precision computation. Second, lower-precision computation consumes less memory footprint and therefore increases the system throughput by allowing the execution of more jobs simultaneously. Third, lower-precision representation reduces the communication cost when exchanging intermediate statistics among the workers is necessary. Saving communication can accelerate the overall training speed. Note that although this book focuses on the optimization of communication, the improvement in computation speed is also significant.

Without loss of generalization, we formalize the representation of floating-point as $(sign, exp, frac)$. The sign term $sign$ denotes whether the number is positive or negative, the exponent width term exp determines the range of the floating-point format, and the significand term $frac$ represents the digit of number, which also decides the representation precision. For example, Fig. 2.20 showcases the floating-point format of the IEEE 754 standard. It defines a 32-bit floating-point number as having 1 bit of $sign$, 8 bits of exp, and 23 bits of $frac$ [57].

The sign bit represents the sign of the number, as well as the sign of the significand term. The 8-bit exponent term originally ranges from 0 to 255, and adding a bias 127 changes the range from -126 to +127. The values of -127 (all bits are 0) and +128 (all bits are 1) are reserved. The significand term contains 23 fraction bits ranging from the first bit to the 22nd bit.

Assuming that the 32-bit number in the figure is denoted by b and each bit is b_i, this real number can be represented as:

$$value = (-1)^{b_{31}} \times 2^{(b_{30}b_{39}...b_{23})_2 - 127} \times (1.b_{22}b_{21}b_0)_2 \tag{2.6}$$

This can also be represented by replacing bits with the exponent term and significand term:

$$value = (-1)^{sign} \times 2^{e-127} \times \left(1 + \sum_{i=1}^{23} b_{23-i}2^{-i}\right) \tag{2.7}$$

Table 2.1 Different formats of floating-point numbers

Format	Total bits	Sign bit	Exponent bit	Significand bit
Single precision (float32)	32	1	8	23
Double precision (float64)	64	1	11	52
Half-precision (float16)	16	1	5	10

We can obtain the number in the example of Fig. 2.20 by putting the bits into the above equation:

$$value = (-1)^1 \times 2^{123-127} \times (1 + 2^{-1} + 2^{-2}) = (-1) \times 2^{-4} \times 1.75 = -0.109375 \quad (2.8)$$

Similarly, other floating-point formats are listed in Table 2.1, including double precision (float64) and half-precision (float16) [57].

Specifically, the IEEE 754 standard defines the 16-bit half-precision floating-point format, which consists of 1 sign bit, 5 exponent bits, and 10 fractional bits. The exponent part is encoded with a bias of 15, yielding an exponential range of $[-14, 15]$ (two exponent values, 0 and 31, are reserved for special cases). Using fewer bits for the exponent term results in a smaller value range. For instance, the value range of single precision is roughly from 1.2×10^{-38} to 3.4×10^{38}, whereas that of double precision is from 2.2×10^{-308} to 1.8×10^{308}. Likewise, using fewer bits for the significand term results in a less precise representation. The precision of single precision for decimals between 1 and 2 is a fixed interval of 2^{-23} and that of half-precision for decimals between 1 and 2 is a fixed interval of 2^{-10}.

Another representation for lower precision is the fixed-point format. Fixed-point representation is a data type for a real number that has a fixed number of digits after (sometimes also before) the radix point (the decimal point "." in English decimal notation). The fractional values in fixed-point numbers usually use a base of 2 or 10. Figure 2.21 shows a 32-bit fixed-point format that contains a sing bit, k integer bits, and $(31 - k)$ fraction bits. The value of this fixed-point number with a base 2 can be represented as:

$$value = (-1)^{b_{31}} \times \left(\sum_{i=0}^{k} b_{31-k+i} 2^i + \sum_{i=0}^{30-k} b_{30-k-i} 2^{-i-1} \right) \quad (2.9)$$

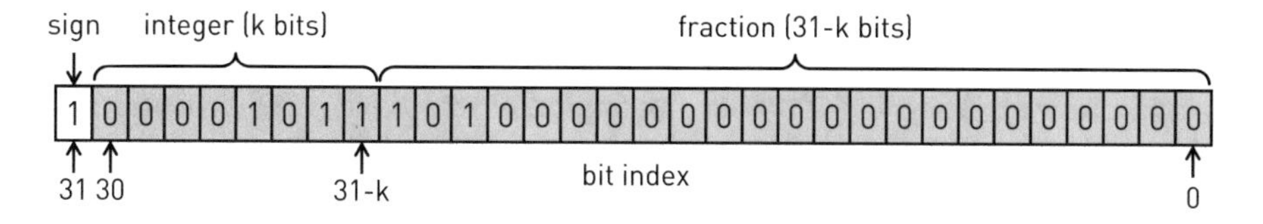

Fig. 2.21 Format of fixed-point number

Based on this equation, the real number in Fig. 2.21 is -11.625. Compared with the floating-point format, the fixed-point format is more suitable for processors that have no FPU (floating-point unit), such as energy-efficient embedded microprocessors and micro-controllers. Generally, the floating-point format offers a higher precision than the fixed-point format because the floating-point format can represent extremely small or extremely large numbers while fixed-point has a limitation on the number of digits.

Many existing works have explored lower precision in training machine learning models [58].

- The earliest work can trace back to the 1980s. Iwata et al design Neuro Turbo [59], an accelerator for neural networks using a 24-bit floating-point digital signal processor.
- Hammerstrom [60] designs a VLSI architecture to perform on-chip model training which supports 8-bit lower-precision representation.
- Holt et al. [61] conduct theoretical analysis on the error brought by low-precision fixed-point numbers in training neural networks. Their results show that 13–15 bits of weight precision and 8–9 bits of activation precision are sufficient for successful forward retrieval and back-propagation learning in a multilayer perceptron model.
- Gupta et al. [62] study the effect of limited precision data representation and computation on neural network training. They observe that the rounding mode adopted when converting a number into a lower-precision fixed-point representation plays a crucial role in determining the network's behavior during training. Through an empirical study of different rounding schemes, such as round-to-nearest and stochastic rounding, they find that conventional rounding schemes fail while adopting stochastic rounding during deep neural network training delivers results nearly identical to 32-bit floating-point computations.
- Courbariaux et al. [63] state that multipliers are the most space and power-expensive arithmetic operators of the digital implementation of deep neural networks. They propose training deep neural networks with low-precision multipliers and high precision accumulators and using a higher precision for the parameters during the updates. The evaluation results find that very low precision is sufficient for training neural networks, e.g., it is possible to train Maxout networks with 10 bits multiplications.
- Vanhoucke et al. [64] improve the training speed of neural networks on CPUs by leveraging SSSE3 and SSE4 low-precision instructions. They use the Intel SSSE3 instruction set to take 16-bit representations for activations and weights.
- Savich et al. [65] study the impact of arithmetic representations for multilayer perceptrons trained using error back-propagation on FPGAs (field-programmable gate arrays). They include both floating-point and fixed-point formats, with considerations of representation precision and FPGA infrastructure. They find that the resource requirements of networks implemented with fixed-point arithmetic are approximately two times less than their counterparts at the floating

point, with similar precision and range of the data representation used, without compromising the effectiveness of training and operational function.

- Hwang et al. [66] optimize the back-propagation based retraining with fixed-point representation. The designed fixed-point networks use a 3-bit signal for ternary weights (+1, 0, and −1), showing only negligible performance loss when compared to the floating-point counterparts. The back-propagation for retraining uses quantized weights and fixed-point signals to compute the output but utilizes high precision values for adapting the networks. They also find that the performance gap between the floating-point and fixed-point networks shrinks as the number of units in each layer increases.
- Flexpoint data format [67] is proposed to replace the 32-bit floating-point format in training and inference. Flexpoint calculates a shared exponent for the tensors that is dynamically adjusted to minimize overflows and maximize the available dynamic range. In a simulator implemented with the neon deep learning framework, 16-bit Flexpoint closely matches 32-bit floating-point in training convolutional neural networks and generative adversarial networks, without the tuning of model hyperparameters.
- Kim and Smaragdis [68] propose a bitwise version of artificial neural networks, where all the inputs, weights, biases, hidden units, and outputs can be represented with single bits and operated on using simple bitwise logic. They design Bitwise Neural Network (BNN), which is especially suitable for resource-constrained environments since it replaces either floating or fixed-point arithmetic with significantly more efficient bitwise operations.
- Miyashita et al. [69] find that the weights and activations in a trained network naturally have nonuniform distributions. They thereby propose a new data representation for convolutional neural networks that encodes data to 3 bits with negligible loss in classification performance for convolutional neural networks. Using nonuniform base-2 logarithmic representation to encode weights and activations and perform dot products enables networks, the proposed method achieves higher classification accuracies than fixed points at the same resolution and eliminates bulky digital multipliers.
- XNOR-Net [70] represents both the filters and the input to convolutional layers as a binary format. XNOR-Net approximates convolutions using primarily binary operations. It can reduce the size of the network by 32 times and provide the possibility of loading very deep neural networks into portable devices with limited memory
- Mellempudi1 et al. [71] propose a cluster-based quantization method that exploits local correlations in a dynamic range of the parameters to minimize the impact of quantization on overall accuracy. They constrain the activations to 8-bits thus enabling an 8-bit full integer inference pipeline. On Resnet-101 using 8-bit activations, the error from the best published full-precision (float32) result is within 6% for ternary weights and within 2% for 4-bit weights.
- Micikevicius [72] introduces a methodology for training deep neural networks using half-precision floating-point numbers, without losing model accuracy or having to modify hyperparameters. All the weights, activations, and gradients

are stored in IEEE half-precision format. They further propose techniques for preventing the loss of critical information—(1) maintain a single precision copy of weights that accumulate the gradients after each optimizer step; (2) propose loss-scaling to preserve gradient values with small magnitudes; (3) use half-precision arithmetic that accumulates into single precision outputs, which are converted to half-precision before storing in memory.

- Das et al. [73] focus on Integer Fused-Multiply-and-Accumulate (FMA) operations which take two pairs of INT16 operands and accumulate results into an INT32 output. They propose a shared exponent representation of tensors and develop a Dynamic Fixed Point (DFP) scheme suitable for common neural network operations. The performance of this method is demonstrated with the implementation of CNN training for ResNet-50, GoogLeNet-v1, VGG-16, and AlexNet; training these networks with mixed-precision DFP16 for ImageNet-1K classification task.
- Wang et al. [74] devise an 8-bit floating-point format (*float8*) consisting of 1-bit sign term, 5-bit exponent term and 2-bit significand term. The authors have demonstrated DNN training with the designed 8-bit floating-point numbers achieves 2–4× speedup without compromising the accuracy. A problem is that the reduced-precision additions (used in partial product accumulations and weight updates) can result in swamping errors causing accuracy degradation during training. To minimize this error, they propose two new techniques, chunk-based accumulation, and floating-point stochastic rounding that enable a reduction of bit-precision for additions down to 16 bits—as well as implement them in hardware. Across a wide spectrum of popular DNN benchmarks and datasets, this mixed-precision float-8 training technique achieves the same accuracy levels as the float-32 baseline.

If these lower-precision methods are applied in distributed training, they can save significant communication overhead compared with traditional training with single precision or double precision.

In addition to these algorithmic contributions, many hardware provide fundamental infrastructure for the implementation of lower-precision format, and popular training engines provide a delicately designed interface for lower-precision computation. Single precision (`float32`) and double precision (`float64`) formats are naturally supported by most hardware (e.g., x86 CPUs and NVIDIA/AMD GPUs) and most computation software. For lower-precision representations, they are not supported in x86 CPUs and poorly supported in earlier GPUs. While since the trend towards using half-precision instead of single precision in training machine learning models, there appears a mixed-precision training mode that uses both single- and half-precision representations. The hardware providers have designed modern GPUs that originally support lower precision, such as the NVIDIA RTX series. NVIDIA also develops programming tools available for mixed-precision computing since Pascal architecture and CUDA 8. The Volta generation of NVIDIA GPUs introduces Tensor Cores, which provide more throughput than single precision executions [75]. For instance, the V100 GPU has 640 Tensor Cores, so they can perform 4×4 multiplications all at the same time. The theoretical peak performance of the Tensor Cores on the V100 is approximately 120 TFLOPS—an order of magnitude (10x) faster than double precision (float64) and approximately 4 times faster than single precision (float32). For practitioners, the most popular computation engines, e.g., TensorFlow, PyTorch, MXNet, and NumPy, have offered easy-to-use programming interfaces for half-precision (float16).

2.5.2 *Communication Compression*

Lower-precision representations can reduce memory footprint, accelerate computation, and decrease the overhead on communication at the expense of introduced errors during computation. To optimize communication, which is the focus of this section, an alternative is to compress the intermediate statistics before they are transmitted.

Data compression is a classical problem in computer science and information theory. The goal is to encode (or compress) information using fewer bits than the original data representation and preserve data information after decoding (or decompressing). Data compression is valuable for cases where storage and bandwidth are expensive or the bottleneck, with extra computational costs during compression and decompression. Regarding the compressed data type, the compression algorithms can be categorized into methods for integer numbers and methods for floating-point numbers. There are many mature data compression algorithms for integer numbers, such as RLE (Run-length Encoding), Huffman, and Rice, which can accurately encode and decode data. However, these approaches for integer numbers cannot be applied for many machine learning training scenarios since the model parameters and intermediate statistics are mostly floating-point numbers.

According to the compression performance, there is lossless compression and lossy compression. Lossless compression reduces the data size by eliminating statistical redundancy, meanwhile without losing any information. In contrast, lossy compression shrinks the data size by removing both unnecessary and less important information while risking the loss of information after decompression.

2.5.2.1 Lossless Compression for Integer Numbers

This section describes several traditional lossless compression methods for integer numbers. This section is necessary as some compression algorithms designed for machine learning training leverage the compression of integer numbers.

RLE

Run-length Encoding [76], RLE in short, is a lossless compression scheme when the same data item may occur consecutively, e.g., integer numbers or alphabets. The basic idea is to replace consecutive items with the frequency of occurrence. For example, RLE can be used for the compression of images, where many adjacent pixels are the same. Assuming the original data contains a segment of consecutive repeated items $AAAAABBB$, RLE encodes them as the combination of occurrence and data item—$5A3B$. If each item needs one byte, RLE reduces the data size from 8 bytes to 4 bytes.

Huffman

Huffman coding is a widely adopted prefix code standard for lossless compression in computer science and information theory, designed by David Huffman in 1952 [77]. Huffman coding converts each item (e.g., a character in a file) in the original data to a variable-length code. Each variable-length code is generated according to the estimated probability or occurrence frequency of each item. The intuition is using shorter codes for high-frequency items and longer codes for low-frequency items. In this way, the expected average length after Huffman coding is minimized.

For example, according to statistics, e is the most frequent character in English articles and z is the least frequent. When using Huffman coding to compress an English article, e is probably to be represented as 1 bit, while z is likely to be encoded with 25 bits. Compared with the original format, which represents each character with one byte (8 bits), e consumes only $\frac{1}{8}$ of the original space, but z needs 3 times more space. If given an accurate prior estimation for the occurrence probabilities of the characters, RLE can obtain a significantly high compression ratio.

The fundamental idea of RLE is to build a binary tree for a sequence of symbols, and the size of the tree is decided by the number of individual symbols. A binary tree in RLE contains leaf nodes and internal nodes. Each leaf node represents a unique symbol and its weights (appearance frequency). Each internal node has two child nodes, and its weight summarizes the weights of two child nodes. Starting from the root node, each split of the internal node is given a bit—a bit "0" for the left path (left child) and a bit "1" for the right path (right child).

For an input sequence of symbols, the procedure of RLE is as follows:

1. Calculate the occurrence frequency and probability for each symbol.
2. Create a leaf node for each symbol, and use the probability as the attribute of the tree. All the leaf nodes are put into a priority queue.
3. While there is more than one node in the queue.
    a. Select and remove the two nodes of the highest priority from the queue. A node that has a low probability is considered a high priority.
    b. Create a new internal node with these two nodes as its child nodes. The probability of the new internal node is the sum of two child nodes' probabilities.
    c. Add the new internal node to the queue.
4. The only remaining node is the root node. The tree building process is finished.

Figure 2.22 gives an example of RLE. There are five different symbols, i.e., A, B, C, D, E, in the input sequence. The table lists the counts and probabilities of these symbols. The building of the Huffman tree contains the following steps.

Character	A	B	C	D	E
Count	15	7	6	6	5
Probability	0.385	0.179	0.154	0.154	0.128
Code	0	100	101	110	111

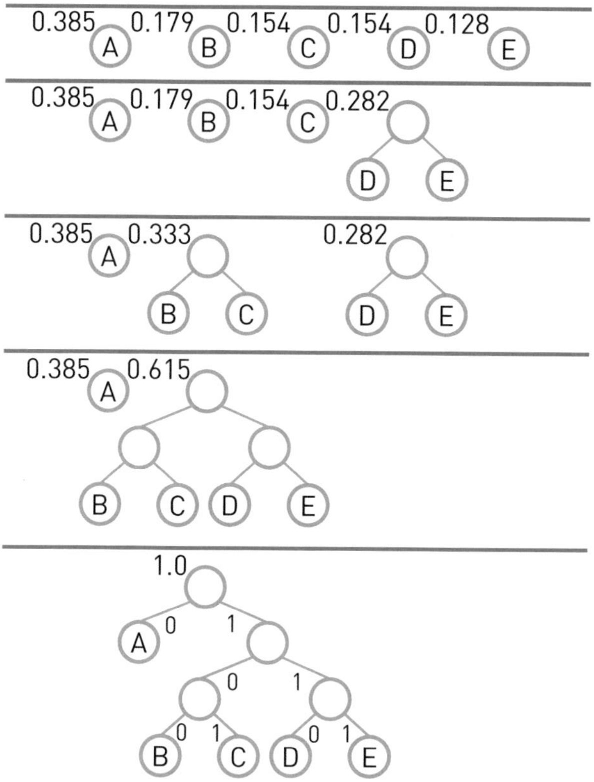

Fig. 2.22 An example of Huffman coding

1. Put five symbols, along with their probabilities, into the queue.
2. D and E are selected and removed from the queue because their probabilities are the two lowest. Create an internal node DE, with D as its left child node (bit 0) and E as its right child node (bit 1). The probability of the new internal node is 0.282, the sum of D and E. The new internal node DE is added to the queue.
3. B and C are then selected and removed from the queue. Create an internal node BC, whose left child node is B and right child node is C. BC is added to the queue with a probability of 0.333.
4. BC and CD are selected, forming a new internal node BCDE.
5. Finally, A and BCDE are merged so that the building process is finished.

The height of the Huffman tree is 4, and all symbols are on leaf nodes. The coding length of A is 1 (bit 0), whereas the coding lengths of other symbols are 3. This resonates with the purpose of RLE—representing highly frequent symbols with shorter codes.

2.5.2.2 Lossless Compression for Sparse Matrices

In the training process of machine learning models, the involved data, including the training data, model parameters, and intermediate statistics, are normally stored in the form of matrices or vectors. An important characteristic of many large-scale datasets is the data sparsity caused by the large feature space, that is, many elements in the raw data matrices/vectors are zero. If all the elements in the training data are stored, there will be many redundancies in the data representation. Some formats of data storage provide space-efficient optimizations for sparse matrices/vectors, which can compress data representation in machine learning workloads without loss. This section describes three popular data formats—COO, CSR, and CSC.

COO

Coordinate (COO) is the simplest data storage format for a sparse matrix. Every nonzero element is represented by a triple—(*row*,*col*,*value*). *row* denotes the row index of the element, *col* denotes the column index, and *value* denotes the element value. As shown in Fig. 2.23, COO stores these triples in three arrays, i.e., the row index array, the column index array, and the value array.

CSR

Compressed Sparse Row (CSR) is a more efficient storage format for sparse matrix. CSR also stores three types of data—values, column indices, and row offsets. Different from COO, CSR generates a compressed format for the whole matrix rather than compressing each nonzero element with a triple.

The column index array and value array, which store the column indices and values of nonzero elements, are the same as COO. The row offset array stores an offset for every row of the matrix. Specifically, for the first nonzero element in each row, the corresponding row offset is its index in the value array. For example, in

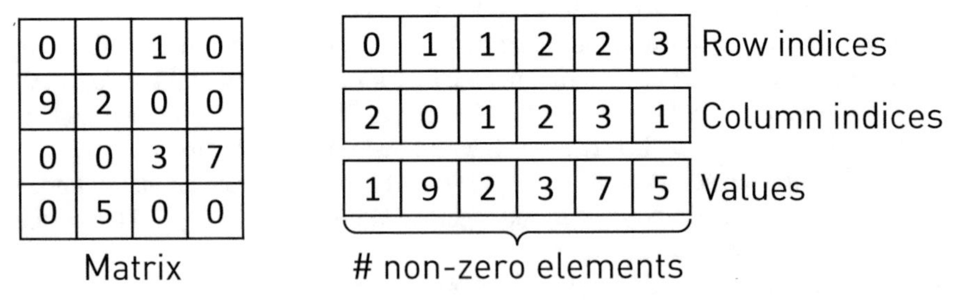

Fig. 2.23 COO format

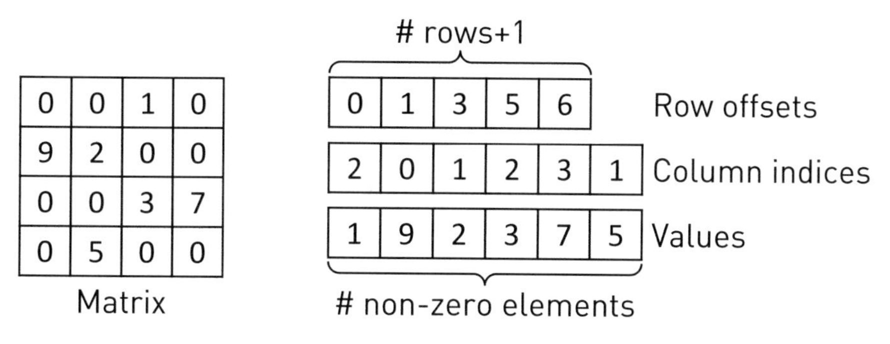

Fig. 2.24 CSR format

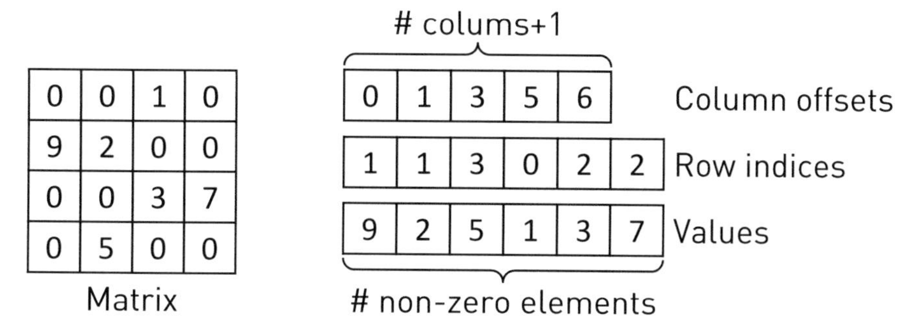

Fig. 2.25 CSC format

Fig. 2.24, the first nonzero element in the first row of the matrix is 1, and its index in the value array is 0, which is treated as its row offset. Likewise, the row offset of element 9 is 1, the row offset of element 3 is 3, and the row offset of element 5 is 5. An extra offset 6, the total number of nonzero elements, is added to the row offset array.

CSR can significantly reduce the storage space. According to a study [78], CSR format costs 8.5 bytes on average for every nonzero 32-bit floating-point element, and 12.5 bytes for every 64-bit floating-point element. However, CSR needs more computation costs than COO due to its complex coding scheme.

CSC

Compressed Sparse Column (CSC) is a column-first sparse storage format. CSC is similar to CSR except that values are stored by column. Different from CSR format, the value array of CSC in Fig. 2.25 is {9, 2, 5, 1, 3, 7}. A row index is stored for each value, and column offsets are stored to indicate the starting index of each column in the value array.

2.5.2.3 Lossy Compression for Floating-point Numbers

The last section introduces methods that can be used for compressing floating-point matrices without any loss. However, for many gradient-based optimization algorithms, the transmitted floating-point intermediate statistics (generally gradients) are dense. The methods designed for compressing sparse matrices are not useful. To reduce the communication overheads in training gradient optimization algorithms, researchers explore lossy compression approaches that represent gradients with a smaller size at the expense of information loss during compression. There are two lines in the area—gradient quantization which represents each gradient dimension with smaller size and gradient sparsification which reduces the number of gradient dimensions.

Gradient Quantization

Gradient quantization is a class of algorithms that leverage a quantification strategy to transform a floating-point number to an integer according to the value range of original data.

Wen et al. [79] propose TernGrad, which quantizes gradients to three ternary levels to reduce the overhead of gradient synchronization. Detailedly, TernGrad randomly quantizes a gradient vector g to a ternary vector with value $\in \{-1, 0, +1\}$:

$$\begin{aligned} \tilde{g} &= s \cdot sign(g) \circ b \\ s &= max(abs(g)) = ||g||_\infty \end{aligned} \tag{2.10}$$

where s is a scalar term that is the maximum absolute norm of g, $\circ$ is the Hadamard product, and $sign(\cdot)$ is the sign of each item in the vector. b is a random binary vector introduced with a stochastic rounding strategy based on the Bernoulli distribution:

$$P(b_k = 1) = \frac{|g_k|}{s_k}, \qquad P(b_k = 0) = 1 - \frac{|g_k|}{s_k} \tag{2.11}$$

where g_k and b_k are the k-th elements of b_k and g_k, respectively. This stochastic rounding scheme, instead of the deterministic scheme, can provide an unbiased expectation.

QSGD [54] proposes an intuitive stochastic quantization scheme—given the gradient vector at a processor, it quantizes each component by randomized rounding to a discrete set of values, in a principled way that preserves the statistical properties of the original. The quantization function $Q(v)$ is involved with a hyper-parameter s, indicating the number of quantization levels. Specifically, QSGD defines s uniformly distributed levels between 0 and 1. For an item v_i in a given input vector $\mathbf{v} \neq \mathbf{0}$, Let $0 \leq l < s$ be an integer such that $\frac{v_i}{\mathbf{v}_2} \in [\frac{l}{s}, \frac{l+1}{s}]$. Hence, $[\frac{l}{s}, \frac{l+1}{s}]$ is the quantization interval of $\frac{v_i}{\mathbf{v}_2}$. $\xi_i(\mathbf{v}, s)$ is defined as:

$$\xi_i(\mathbf{v}, s) = \begin{cases} \frac{l}{s} & with\ probability\ 1 - p(\frac{v_i}{||v||}, s); \\ \frac{l+1}{s} & otherwise \end{cases} \tag{2.12}$$

where $p(a, s) = as - l$ for any $a \in [0, 1]$. If $\mathbf{v} = \mathbf{0}$, then $Q(\mathbf{v}, s) = \mathbf{0}$. The distribution of $\xi_i(\mathbf{v}, s)$ has minimal variance over $\xi_i(\mathbf{v}, s)$, $Q(\mathbf{v})$ is defined as:

$$Q(v_i) = ||\mathbf{v}||_2 \cdot sign(v_i) \cdot \xi_i(\mathbf{v}, s) \tag{2.13}$$

The authors prove the unbiasedness of QSGD and give its variance bound. Experiments show that 4bit- or 8bit-QSGD is sufficient to recover or even slightly improve accuracy compared with the full-precision variant on a variety of tasks while ensuring nontrivial speedup.

Researchers from Microsoft Research propose 1-bit SGD that aggressively quantizes the gradients to one bit per value [80]. They show that this does not or almost not reduce accuracies—but only if the quantization error is carried forward across minibatches, i.e., the error in quantizing the gradient in one minibatch is added (fed back) to the gradient of the next minibatch. The authors implement 1-bit SGD with AdaGrad [81] and verify its effectiveness over a production scale model of 160M parameters.

Strom [82] proposes quantizing the gradient and packing both the quantized gradient and the index in a single 32-bit integer field. The quantization error is not discarded but added back to the gradient residual (a.k.a, error feedback). Empirical study finds that 1-bit quantization for the gradient is sufficient and carries no significant degradation in neither accuracy nor convergence speed, leaving 31 bits of address space for the index, which is more than enough for all most use cases.

ZipML [55] focuses on reducing the variance of stochastic gradient quantization. They notice that different methods for setting the quantization points have different variances. Through performing an analysis of the optimal quantizations for various settings, they observe that the uniform quantization approach popularly used by state-of-the-art end-to-end low-precision deep learning training systems when more than 1 bit is used is suboptimal. To address this problem, they formulate an independent optimization problem to select optimal quantization points and solve it optimally with an efficient dynamic programming algorithm that only needs to scan the data in a single pass.

Similarly, Jiang et al. also find that uniform quantization introduces a large variance for nonuniformly distributed gradients [83]. They propose a nonuniform quantization method, called SketchML, that uses a quantile data sketch [84–86] to fit the distribution of gradients and generates nonuniform quantization points. SketchML supports sparse gradients by extending a frequency sketch [87].

TinyScript [88] adopts a nonuniform quantization framework for training DNNs. TINYSCRIPT models the values with a family of Weibull distributions to compute optimal quantization points and leverages the distribution property to achieve minimum quantization variance. Empirical results show that TINYSCRIPT obtains lower quantization variance than the uniform-based mechanism, and therefore achieves a higher compression rate with model accuracy on par with full-precision training.

Gradient Sparsification

Another line of research focusing on compressing floating-point gradients is performing sparsification to the gradient vector by only sending the important coordinates in the gradient vector and dropping the less important ones.

Aji and Heafield [89] make distributed stochastic gradient descent (SGD) faster by exchanging sparse updates instead of dense updates. Considering gradient updates are skewed as most updates are near zero, so the proposed method maps the 99% smallest updates (by absolute value) to zero then exchange sparse matrices, reducing the communication size to 50× smaller with coordinate-value encoding.

Stich [90] analyses the theoretical convergence of the in-memory stochastic gradient descent (SGD) with top-k or random-k sparsification-based compression and finds that this scheme converges at the same rate as vanilla SGD when equipped with error compensation (keeping track of accumulated errors in memory). The experiments verify the drastic reduction in communication cost by demonstrating that SGD with gradient sparsification requires one to two orders of magnitude fewer bits to be communicated than QSGD [54] while converging to the same accuracy. Besides, top-k sparsification shows an advantage over random sparsification in the serial setting but not in the multicore shared-memory implementation.

Wang et al. [91] find that gradient optimization algorithms involve many atomic decompositions, such as the entry-wise decomposition of gradient $g = \sum_i g_i e_i$ and the singular value decomposition for matrices. Based on this insight, they present ATOMO, a general sparsification method for distributed stochastic gradient optimization algorithms. Given a gradient, an atomic decomposition, and a sparsity budget, ATOMO gives a random unbiased sparsification of the atoms minimizing variance. They show that recent methods such as QSGD and TernGrad are special cases of ATOMO. They focus on the use of ATOMO for sparsifying matrices, especially the gradients in neural network training. We show that applying ATOMO to the singular values of these matrices can lead to faster training than both vanilla SGD or QSGD, for the same communication budget.

Lin et al. find that 99.9% of the gradient exchange in distributed SGD is redundant. They propose Deep Gradient Compression (DGC) to greatly reduce communication cost [92]. To preserve accuracy during this compression, DGC employs four techniques: momentum correction, local gradient clipping, momentum factor masking, and warm-up training. Compared to the previous work, DGC pushes the gradient compression ratio to up to 600× for the whole model (same compression ratio for all layers). Furthermore, DGC does not require extra layer normalization and thus does not need to change the model structure.

References

1. Zinkevich, Martin and Weimer, Markus and Li, Lihong and Smola, Alex J: Parallelized stochastic gradient descent. Advances in Neural Information Processing Systems. 2595–2603 (2010)
2. Jiang, Jiawei and Cui, Bin and Zhang, Ce and Yu, Lele: Heterogeneity-aware distributed parameter servers. Proceedings of the 2017 ACM International Conference on Management of Data. 463–478 (2017)
3. Zhang, Zhipeng and Jiang, Jiawei and Wu, Wentao and Zhang, Ce and Yu, Lele and Cui, Bin: Mllib*: Fast training of glms using spark mllib. 2019 IEEE 35th International Conference on Data Engineering (ICDE). 1778–1789 (2019)
4. Recht, Benjamin and Re, Christopher and Wright, Stephen and Niu, Feng: Hogwild: A lock-free approach to parallelizing stochastic gradient descent. Advances in Neural Information Processing Systems. 693–701 (2011)
5. Dean, Jeffrey and Corrado, Greg and Monga, Rajat and Chen, Kai and Devin, Matthieu and Mao, Mark and Ranzato, Marc'aurelio and Senior, Andrew and Tucker, Paul and Yang, Ke and others: Large scale distributed deep networks. Advances in Neural Information Processing Systems. 1223–1231 (2012)
6. Chen, Tianqi and Guestrin, Carlos: Xgboost: A scalable tree boosting system. Proceedings of the 22nd ACM SIGKDD International Conference on Knowledge Discovery and Data Mining. 785–794 (2016)
7. Goyal, Priya and Dollár, Piotr and Girshick, Ross and Noordhuis, Pieter and Wesolowski, Lukasz and Kyrola, Aapo and Tulloch, Andrew and Jia, Yangqing and He, Kaiming: Accurate, large minibatch SGD: Training ImageNet in 1 hour. arXiv preprint arXiv:1706.02677. (2017)
8. He, Kaiming and Zhang, Xiangyu and Ren, Shaoqing and Sun, Jian: Deep residual learning for image recognition. Proceedings of the IEEE Conference on Computer Vision and Pattern Recognition. 770–778 (2016)
9. Deng, Jia and Dong, Wei and Socher, Richard and Li, Li-Jia and Li, Kai and Fei-Fei, Li: ImageNet: A large-scale hierarchical image database. 2009 IEEE Conference on Computer Vision and Pattern Recognition. 248–255 (2009)
10. Zhang, Ce and Ré, Christopher: DimmWitted: a study of main-memory statistical analytics. Proceedings of the VLDB Endowment. 7(12), 1283–1294 (2014)
11. Zhang, Zhipeng and Wu, Wentao and Jiang, Jiawei and Yu, Lele and Cui, Bin and Zhang, Ce: ColumnSGD: A Column-oriented Framework for Distributed Stochastic Gradient Descent. 2020 IEEE 36th International Conference on Data Engineering (ICDE). 1513–1524 (2020)
12. Fu, Fangeheng and Jiang, Jiawei and Shao, Yingxia and Cui, Bin: An experimental evaluation of large scale GBDT systems. Proceedings of the VLDB Endowment. 12(11), 1357–1370 (2019)
13. Krizhevsky, Alex and Sutskever, Ilya and Hinton, Geoffrey E: ImageNet classification with deep convolutional neural networks. Communications of the ACM. 60(6), 84–90 (2017)
14. Brown, Tom B and Mann, Benjamin and Ryder, Nick and Subbiah, Melanie and Kaplan, Jared and Dhariwal, Prafulla and Neelakantan, Arvind and Shyam, Pranav and Sastry, Girish and Askell, Amanda and others: Language models are few-shot learners. arXiv preprint arXiv:2005.14165. (2020)
15. Gemulla, Rainer and Nijkamp, Erik and Haas, Peter J and Sismanis, Yannis: Large-scale matrix factorization with distributed stochastic gradient descent. Proceedings of the 17th ACM SIGKDD International Conference on Knowledge Discovery and Data Mining. 69–77 (2011)
16. Wang, Wei and Chen, Gang and Dinh, Anh Tien Tuan and Gao, Jinyang and Ooi, Beng Chin and Tan, Kian-Lee and Wang, Sheng: SINGA: Putting deep learning in the hands of multimedia users. Proceedings of the 23rd ACM International Conference on Multimedia. 25–34 (2015)

17. Jiang, Jiawei and Cui, Bin and Zhang, Ce and Fu, Fangcheng: Dimboost: Boosting gradient boosting decision tree to higher dimensions. Proceedings of the 2018 International Conference on Management of Data. 1363–1376 (2018)
18. Walker, David W and Dongarra, Jack J: MPI: a standard message passing interface. Supercomputer. 12, 56–68 (1996)
19. Gropp, William and Gropp, William D and Lusk, Ewing and Skjellum, Anthony and Lusk, Argonne Distinguished Fellow Emeritus Ewing: Using MPI: portable parallel programming with the message-passing interface. MIT press. 1 (1999)
20. Thakur, Rajeev and Rabenseifner, Rolf and Gropp, William: Optimization of collective communication operations in MPICH. The International Journal of High Performance Computing Applications. 19(1), 49–66 (2005)
21. Open MPI, https://www.open-mpi.org/
22. MPICH, https://www.mpich.org/
23. MPI4PY, https://mpi4py.readthedocs.io/en/stable/
24. Nelson, Bruce Jay: REMOTE PROCEDURE CALL. (1982)
25. Bershad, Brian and Anderson, Thomas and Lazowska, Edward and Levy, Henry: Lightweight remote procedure call. ACM SIGOPS Operating Systems Review. 23(5), 102–113 (1989)
26. Li ,Jinyang: Distributed Systems, http://www.news.cs.nyu.edu/~jinyang/fa09/ (2020)
27. gRPC, https://grpc.io/
28. Apache Thrift, https://thrift.apache.org/
29. Apache Dubbo, http://dubbo.apache.org/
30. Dean, Jeffrey and Ghemawat, Sanjay: MapReduce: simplified data processing on large clusters. Communications of the ACM. 51(1), 107–113 (2008)
31. Chu, Cheng-Tao and Kim, Sang K and Lin, Yi-An and Yu, YuanYuan and Bradski, Gary and Olukotun, Kunle and Ng, Andrew Y: Map-reduce for machine learning on multicore. Advances in Neural Information Processing Systems. 281–288 (2007)
32. Apache Mahout, http://mahout.apache.org/
33. Ghoting, Amol and Krishnamurthy, Rajasekar and Pednault, Edwin and Reinwald, Berthold and Sindhwani, Vikas and Tatikonda, Shirish and Tian, Yuanyuan and Vaithyanathan, Shivakumar: SystemML: Declarative machine learning on MapReduce. 2011 IEEE 27th International Conference on Data Engineering. 231–242 (2011)
34. Apache Hadoop, https://hadoop.apache.org/
35. Zaharia, Matei and Xin, Reynold S and Wendell, Patrick and Das, Tathagata and Armbrust, Michael and Dave, Ankur and Meng, Xiangrui and Rosen, Josh and Venkataraman, Shivaram and Franklin, Michael J and others: Apache spark: a unified engine for big data processing. Communications of the ACM. 59(11), 56–65 (2016)
36. Zaharia, Matei and Chowdhury, Mosharaf and Das, Tathagata and Dave, Ankur and Ma, Justin and McCauly, Murphy and Franklin, Michael J and Shenker, Scott and Stoica, Ion: Resilient distributed datasets: A fault-tolerant abstraction for in-memory cluster computing. Presented as part of the 9th USENIX Symposium on Networked Systems Design and Implementation (NSDI 12). 15–28 (2012)
37. Meng, Xiangrui and Bradley, Joseph and Yavuz, Burak and Sparks, Evan and Venkataraman, Shivaram and Liu, Davies and Freeman, Jeremy and Tsai, DB and Amde, Manish and Owen, Sean and others: Mllib: Machine learning in Apache Spark. The Journal of Machine Learning Research. 17(1), 1235–1241 (2016)
38. Chang, Chih-Chung and Lin, Chih-Jen: LIBSVM: A library for support vector machines. ACM transactions on intelligent systems and technology (TIST). 2(3), 1–27 (2011)
39. xlearn, https://github.com/aksnzhy/xlearn
40. Nitzberg, Bill and Lo, Virginia: Distributed shared memory: A survey of issues and algorithms. Computer. 24(8), 52–60 (1991)
41. Jiang, Jie and Yu, Lele and Jiang, Jiawei and Liu, Yuhong and Cui, Bin: Angel: a new large-scale machine learning system. National Science Review. 5(2), 216–236 (2018)
42. Smola, Alexander and Narayanamurthy, Shravan: An architecture for parallel topic models. Proceedings of the VLDB Endowment. 3(1-2), 703–710 (2010)

43. Xing, Eric P and Ho, Qirong and Dai, Wei and Kim, Jin Kyu and Wei, Jinliang and Lee, Seunghak and Zheng, Xun and Xie, Pengtao and Kumar, Abhimanu and Yu, Yaoliang: Petuum: A new platform for distributed machine learning on big data. IEEE Transactions on Big Data. 1(2), 49–67 (2015)
44. Cui, Henggang and Cipar, James and Ho, Qirong and Kim, Jin Kyu and Lee, Seunghak and Kumar, Abhimanu and Wei, Jinliang and Dai, Wei and Ganger, Gregory R and Gibbons, Phillip B and others: Exploiting bounded staleness to speed up big data analytics. 2014 USENIX Annual Technical Conference (USENIX ATC 14). 37–48 (2014)
45. McMahan, Brendan and Moore, Eider and Ramage, Daniel and Hampson, Seth and y Arcas, Blaise Aguera: Communication-efficient learning of deep networks from decentralized data. Artificial Intelligence and Statistics. 1273–1282 (2017)
46. Abadi, Martín and Barham, Paul and Chen, Jianmin and Chen, Zhifeng and Davis, Andy and Dean, Jeffrey and Devin, Matthieu and Ghemawat, Sanjay and Irving, Geoffrey and Isard, Michael and others: TensorFlow: A system for large-scale machine learning. 12th USENIX Symposium on Operating Systems Design and Implementation (OSDI 16). 265–283 (2016)
47. Paszke, Adam and Gross, Sam and Massa, Francisco and Lerer, Adam and Bradbury, James and Chanan, Gregory and Killeen, Trevor and Lin, Zeming and Gimelshein, Natalia and Antiga, Luca and others: PyTorch: An imperative style, high-performance deep learning library. Advances in Neural Information Processing Systems. 8026–8037 (2019)
48. Li, Mu and Andersen, David G and Smola, Alexander: Distributed delayed proximal gradient methods. NIPS Workshop on Optimization for Machine Learning. 3, 3 (2013)
49. Agarwal, Alekh and Duchi, John C: Distributed delayed stochastic optimization. Advances in Neural Information Processing Systems. 873–881 (2011)
50. Zhang, Ruiliang and Zheng, Shuai and Kwok, James T: Asynchronous distributed semi-stochastic gradient optimization. arXiv preprint arXiv:1508.01633. (2015)
51. Zinkevich, Martin and Langford, John and Smola, Alex: Slow learners are fast. Advances in Neural Information Processing Systems. 22, 2331–2339 (2009)
52. Ho, Qirong and Cipar, James and Cui, Henggang and Lee, Seunghak and Kim, Jin Kyu and Gibbons, Phillip B and Gibson, Garth A and Ganger, Greg and Xing, Eric P: More effective distributed ml via a stale synchronous parallel parameter server. Advances in Neural Information Processing Systems. 1223–1231 (2013)
53. Dai, Wei and Kumar, Abhimanu and Wei, Jinliang and Ho, Qirong and Gibson, Garth and Xing, Eric P: High-performance distributed ML at scale through parameter server consistency models. arXiv preprint arXiv:1410.8043. (2014)
54. Alistarh, Dan and Grubic, Demjan and Li, Jerry and Tomioka, Ryota and Vojnovic, Milan: QSGD: Communication-efficient SGD via gradient quantization and encoding. Advances in Neural Information Processing Systems. 1709–1720 (2017)
55. Zhang, Hantian and Li, Jerry and Kara, Kaan and Alistarh, Dan and Liu, Ji and Zhang, Ce: ZipML: Training linear models with end-to-end low precision, and a little bit of deep learning. International Conference on Machine Learning. 4035–4043 (2017)
56. Audhkhasi, Kartik and Osoba, Osonde and Kosko, Bart: Noise benefits in backpropagation and deep bidirectional pre-training. The 2013 International Joint Conference on Neural Networks (IJCNN). 1–8 (2013)
57. Single precision floating-point format, https://en.wikipedia.org/wiki/Single-precision_floating-point_format
58. Rodriguez, Andres and Segal, Eden and Meiri, Etay and Fomenko, Evarist and Kim, Y Jim and Shen, Haihao and Ziv, Barukh: Lower numerical precision deep learning inference and training. Intel White Paper. 3 (2018)
59. Iwata, Akira and Yoshida, Yukio and Matsuda, Satoshi and Sato, Yukimasa and Suzumura, Nobuo: An artificial neural network accelerator using general purpose 24 bits floating point digital signal processors. Neural Networks. 171–175 (1989)
60. Hammerstrom, Dan: A VLSI architecture for high-performance, low-cost, on-chip learning. 1990 IJCNN International Joint Conference on Neural Networks. 537–544 (1990)

61. Holi, Jordan L and Hwang, J-N: Finite precision error analysis of neural network hardware implementations. IEEE Transactions on Computers. 42(3), 281–290 (1993)
62. Gupta, Suyog and Agrawal, Ankur and Gopalakrishnan, Kailash and Narayanan, Pritish: Deep learning with limited numerical precision. International Conference on Machine Learning. 1737–1746 (2015)
63. Courbariaux, Matthieu and Bengio, Yoshua and David, Jean-Pierre: Training deep neural networks with low precision multiplications. arXiv preprint arXiv:1412.7024. (2014)
64. Vanhoucke, Vincent and Senior, Andrew and Mao, Mark Z: Improving the speed of neural networks on CPUs. (2011)
65. Savich, Antony W and Moussa, Medhat and Areibi, Shawki: The impact of arithmetic representation on implementing MLP-BP on FPGAs: A study. IEEE Transactions on Neural Networks. 18(1), 240–252 (2007)
66. Hwang, Kyuyeon and Sung, Wonyong: Fixed-point feedforward deep neural network design using weights +1, 0, and -1. 2014 IEEE Workshop on Signal Processing Systems (SiPS). 1–6 (2014)
67. Köster, Urs and Webb, Tristan and Wang, Xin and Nassar, Marcel and Bansal, Arjun K and Constable, William and Elibol, Oguz and Gray, Scott and Hall, Stewart and Hornof, Luke and others: Flexpoint: An adaptive numerical format for efficient training of deep neural networks. Advances in Neural Information Processing Systems. 1742–1752 (2017)
68. Kim, Minje and Smaragdis, Paris: Bitwise neural networks. arXiv preprint arXiv:1601.06071. (2016)
69. Miyashita, Daisuke and Lee, Edward H and Murmann, Boris: Convolutional neural networks using logarithmic data representation. arXiv preprint arXiv:1603.01025. (2016)
70. Rastegari, Mohammad and Ordonez, Vicente and Redmon, Joseph and Farhadi, Ali: Xnor-net: ImageNet classification using binary convolutional neural networks. European Conference on Computer Vision. 525–542 (2016)
71. Mellempudi, Naveen and Kundu, Abhisek and Das, Dipankar and Mudigere, Dheevatsa and Kaul, Bharat: Mixed low-precision deep learning inference using dynamic fixed point. arXiv preprint arXiv:1701.08978. (2017)
72. Micikevicius, Paulius and Narang, Sharan and Alben, Jonah and Diamos, Gregory and Elsen, Erich and Garcia, David and Ginsburg, Boris and Houston, Michael and Kuchaiev, Oleksii and Venkatesh, Ganesh and others: Mixed precision training. arXiv preprint arXiv:1710.03740. (2017)
73. Das, Dipankar and Mellempudi, Naveen and Mudigere, Dheevatsa and Kalamkar, Dhiraj and Avancha, Sasikanth and Banerjee, Kunal and Sridharan, Srinivas and Vaidyanathan, Karthik and Kaul, Bharat and Georganas, Evangelos and others: Mixed precision training of convolutional neural networks using integer operations. arXiv preprint arXiv:1802.00930. (2018)
74. Wang, Naigang and Choi, Jungwook and Brand, Daniel and Chen, Chia-Yu and Gopalakrishnan, Kailash: Training deep neural networks with 8-bit floating point numbers. Advances in Neural Information Processing Systems. 7675–7684 (2018)
75. Mixed-precision in NVIDIA, https://docs.nvidia.com/deeplearning/performance/mixed-precision-training/index.html
76. Golomb, Solomon: Run-length encodings (Corresp.). IEEE Transactions on Information Theory. 12(3), 399–401 (1966)
77. Huffman, David A: A method for the construction of minimum-redundancy codes. Proceedings of the IRE. 40(9), 1098–1101 (1952)
78. Bell, Nathan and Garland, Michael: Efficient sparse matrix-vector multiplication on CUDA. Nvidia Technical Report NVR-2008-004, Nvidia Corporation. (2008)
79. Wen, Wei and Xu, Cong and Yan, Feng and Wu, Chunpeng and Wang, Yandan and Chen, Yiran and Li, Hai: Terngrad: Ternary gradients to reduce communication in distributed deep learning. Advances in Neural Information Processing Systems. 1509–1519 (2017)

80. Seide, Frank and Fu, Hao and Droppo, Jasha and Li, Gang and Yu, Dong: 1-bit stochastic gradient descent and its application to data-parallel distributed training of speech DNNs. Fifteenth Annual Conference of the International Speech Communication Association. (2014)
81. Duchi, John and Hazan, Elad and Singer, Yoram: Adaptive subgradient methods for online learning and stochastic optimization. Journal of Machine Learning Research. 12(7) (2011)
82. Strom, Nikko: Scalable distributed DNN training using commodity GPU cloud computing. Sixteenth Annual Conference of the International Speech Communication Association. (2015)
83. Jiang, Jiawei and Fu, Fangcheng and Yang, Tong and Cui, Bin: Sketchml: Accelerating distributed machine learning with data sketches. Proceedings of the 2018 International Conference on Management of Data. 1269–1284 (2018)
84. Greenwald, Michael and Khanna, Sanjeev: Space-efficient online computation of quantile summaries. ACM SIGMOD Record. 30(2), 58–66 (2001)
85. Zhang, Qi and Wang, Wei: A fast algorithm for approximate quantiles in high speed data streams. 19th International Conference on Scientific and Statistical Database Management (SSDBM 2007). 29–29 (2007)
86. Data Sketches, https://datasketches.github.io/
87. Cormode, Graham and Muthukrishnan, Shan: An improved data stream summary: the count-min sketch and its applications. Journal of Algorithms. 55(1), 258–75 (2005)
88. Fu, Fangcheng and Hu, Yuzheng and He, Yihan and Jiang, Jiawei and Shao, Yingxia and Zhang, Ce and Cui, Bin: Don't waste your bits! squeeze activations and gradients for deep neural networks via TINYSCRIPT. International Conference on Machine Learning. 3304–3314 (2020)
89. Aji, Alham Fikri and Heafield, Kenneth: Sparse communication for distributed gradient descent. arXiv preprint arXiv:1704.05021. (2017)
90. Stich, Sebastian U and Cordonnier, Jean-Baptiste and Jaggi, Martin: Sparsified SGD with memory. Advances in Neural Information Processing Systems. 4447–4458 (2018)
91. Wang, Hongyi and Sievert, Scott and Liu, Shengchao and Charles, Zachary and Papailiopoulos, Dimitris and Wright, Stephen: Atomo: Communication-efficient learning via atomic sparsification. Advances in Neural Information Processing Systems. 9850–9861 (2018)
92. Lin, Yujun and Han, Song and Mao, Huizi and Wang, Yu and Dally, William J: Deep gradient compression: Reducing the communication bandwidth for distributed training. arXiv preprint arXiv:1712.01887. (2017)

Chapter 3
Distributed Gradient Optimization Algorithms

Abstract In this chapter, we will elaborate on state-of-the-art gradient optimization algorithms designed for distributed training of machine learning models. We classify these methods by their targeted machine learning models, including but not limited to generalized linear models, deep learning models, and tree models. For each category of machine learning model, we briefly describe their concepts and principles and then introduce the existing gradient optimization algorithms that can be adopted.

3.1 Linear Models

Generalized linear models are a class of machine learning models popularized by McCullagh and Nelder [1], which are formulated by unifying various statistical models, including linear regression, logistic regression, lasso, and softmax regression.

In general, GLMs model the correlation between the variables x (a.k.a. covariates, influencing variables, or explanatory variables) and the response variable y (a.k.a. dependent variable), in a linear way, although the real relation may not linear. In a generalized linear model, there are three major components:

- *A probability distribution* that models the conditional distribution of the response variable y in terms of the covariate x. The distribution is often from an exponential family, such as Gaussian (Normal), binomial, Poisson, and gamma distributions.
- *A linear predictor* η that generates a prediction with the covariates x in a linear form.
- *A link function* $g(\cdot)$ that transforms the expectation of the response variable $\mu = E(y)$ to the linear predictor. Typical link functions include the identity function and log function.

J. Jiang et al., *Distributed Machine Learning and Gradient Optimization*, Big Data Management, https://doi.org/10.1007/978-981-16-3420-8_3

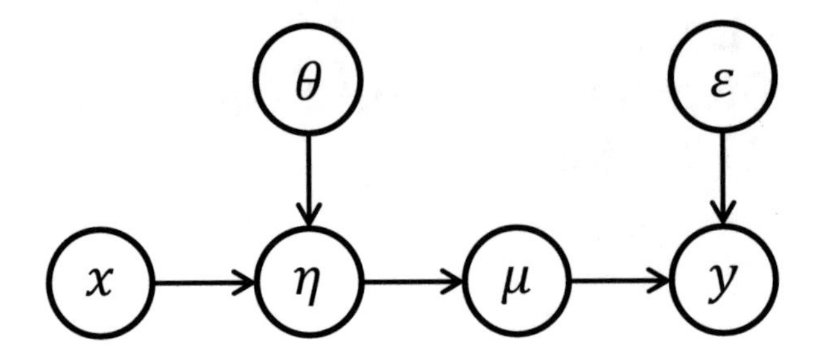

Fig. 3.1 Illustration of generalized linear model

The relationship between the main components is illustrated in Fig. 3.1. Concretely, η is assumed to be a linear combination of the covariate x, and the coefficient β needs to be estimated:

$$\eta = \alpha + x^T\beta = \alpha + \beta_1 x_1 + \beta_2 x_2 + \ldots + \beta_d x_d \tag{3.1}$$

where α is a bias and d is the dimension of the covariate x. The link function defines how the mean (expectation of the response) is calculated with the linear predictor:

$$g(\mu) = \eta = \alpha + x^T\beta \tag{3.2}$$

Since the link function is invertible, the link function can also be represented as

$$\mu = g^{-1}(\eta) = g^{-1}(\alpha + x^T\beta) \tag{3.3}$$

With the definition of the link function, GLM can be seen as a linear model for the specific transformation of the response variable. Table 3.1 shows the commonly used link functions.

The response variable y is assumed to follow an exponential family distribution with mean μ and variance σ. For instance, linear regression assumes a normal distribution, while the other possible probabilities include binomial, Poisson, and gamma distributions. The variance is chosen as a suitable function of the mean value—$V(\mu)$, e.g., some constant in linear regression. For example, assume y has a normal distribution with mean μ and variance σ^2.

$$y \sim N(\mu, \sigma^2) \tag{3.4}$$

Table 3.1 Common link functions in generalized linear models

Link function	$g(\mu)$	$g^{-1}(\eta)$
Identify	μ	η
Log	$\log_e \mu$	e^η
Inverse	μ^{-1}	η^{-1}
Logit	$\log_e \frac{\mu}{1-\mu}$	$\frac{1}{1+e^{-\eta}}$
Probit	$\Phi^{-1}(\mu)$	$\Phi(\eta)$

3.1.1 Formalization of Linear Models

In this book, we focus on several popular (generalized) linear models—linear regression, lasso regression, logistic regression, support vector machine, and multinomial regression. The problem settings are formalized by simplifying the definition of generalized linear models:

- *Input.* Each input instance is a vector $x_i \in \mathbb{R}^d, i = 1, \ldots, n$. d is the dimension of input instance (a.k.a. the number of features).
- *Desired output.* For each input instance, the desired output is $y_i \in \mathbb{R}$. For classification tasks, the desired outputs are discrete labels $y_i \in \mathbb{Z}$, while for regression tasks, the desired outputs are continuous values $y_i \in \mathbb{R}$.
- *Model parameter.* The linear model has model parameters, including the coefficient and the bias term. Without loss of generalization, we merge the coefficient and the bias as $\theta \in \mathbb{R}^d$. This can be achieved by adding a new dimension to each input instance and setting the new dimension as 1. Then, the corresponding dimension in θ is the bias.
- *Predicted output.* The fitted model generates the predicted output of the input instance $\hat{y}_i = \theta \cdot x_i + b$.

To find suitable model parameters for the linear models, an optimization problem is formalized as follows:

$$arg \min_{\theta} f(\theta) = \frac{1}{N} \sum_{i=1}^{N} l(y_i, \hat{y}_i) + \Omega(\theta) \tag{3.5}$$

Here, $f(\theta)$ is the objective function, parameterized by the model parameter θ. The objective comprises the loss function $l(y, \hat{y})$ and the regularization term $\Omega(\theta)$. The loss function estimates the difference between the desired output and the predicted output. The regularization term is leveraged to pose a penalty according to the quantity of the coefficient, which requires the coefficient to approach zero. This can prevent the optimization process from overfitting, typically with some too large values in the coefficient.

The above objective function is often minimized by gradient-based optimization algorithms. At the t-th iteration, the partial gradients w.r.t. the current model parameter θ_t are computed:

$$\nabla f(\theta_t) \equiv f'(\theta_t) \equiv g(\theta_t) \equiv \left[\frac{\partial f}{\partial \theta_{t1}}, \frac{\partial f}{\partial \theta_{t2}}, \ldots, \frac{\partial f}{\partial \theta_{td}} \right]^T \tag{3.6}$$

The optimization algorithm updates the state of the model parameter toward the opposite direction of the gradients:

$$\theta_{t+1} = \theta_t - \eta f'(\theta_t) \tag{3.7}$$

Below, we present the formalization of each considered linear model, i.e., the loss functions and the calculation of gradient.

3.1.2 Overview of Popular Linear Models

We present several popular linear models in this section, including linear regression, lasso, logistic regression, multinomial regression, and support vector machine.

Linear Regression

A simple example of the generalized linear model is linear regression, a.k.a. ridge regression. Regarding the definition of the generalized linear model, the link function is the identity function, and the probability distribution is the normal distribution with a constant variance function:

$$\begin{aligned} \eta &= x^T\theta \\ \mu &= \eta \\ \varepsilon &\sim N(0, \sigma^2) \\ y &= \mu + \varepsilon \end{aligned} \tag{3.8}$$

With this definition, the distribution of the response variable is thereby $y \sim N(\mu, \sigma^2)$.

The optimization problem of linear regression is formalized below:

$$\begin{aligned} \hat{y}_i &= \theta^T x_i \\ l(y_i, \hat{y}_i) &= (\hat{y}_i - y_i)^2 \\ f(\theta) &= \frac{1}{N}\sum_{i=1}^{N} l(y_i, \hat{y}_i) + \frac{1}{2}\lambda||\theta||_2 \\ f'(\theta) &= \frac{1}{N}\sum_{i=1}^{N} \frac{\partial l}{\partial \theta} + \lambda\theta = \frac{1}{N}\sum_{i=1}^{N} 2(\hat{y}_i - y_i)x_i + \lambda\theta \end{aligned} \tag{3.9}$$

The loss function of linear regression is the sum of squared loss between the desired outputs and the predicted outputs, which explicitly measures the quality of the prediction. The regularization term is $l2$-norm regularization—the sum of squared coefficients. The calculation of gradients is straightforward by taking the gradients of the loss function and the regularization.

Lasso

The fitted coefficients of linear regression are generally formed as a dense vector, having nonzero values in every dimension. However, some cases demand model interpretability or selecting a subset of important coefficients. For example, when the input covariates contain too many irrelevant dimensions and only a few dimensions have strong relations with the output, it is of significant value to select these important dimensions for a better understanding of the input and explaining the trained model. Linear regression shrinks the magnitude of the coefficients; however, it cannot achieve feature selection and make the model interpretable. Lasso is designed to fulfill the above goals by adopting $l1$-norm regularization instead of $l2$-norm regularization, that is, penalizing the sum of the absolute values of the coefficients [2–4]. This $l1$ penalty forces many coefficients to be zero, yielding a simpler model that neglects many input features for the output.

Figure 3.2 depicts the fundamental difference of lasso regression and ridge regression (linear regression). Assume the coefficients have two dimensions (θ_1 and θ_2), and the elliptical contours in orange are the loss functions for lasso regression and ridge regression. The constraint regions regarding the regularization are plotted as blue rectangle and circle for lasso and ridge regressions, respectively. The finding of the coefficients can be seen as finding the place where the contours hit the constraint regions. Typically, the hitting point of lasso regression is on the axes, and therefore other dimensions of coefficients are ignored. In contrast, the hitting point of ridge regression is not on the axes; hence, the fitted coefficients are all nonzero values.

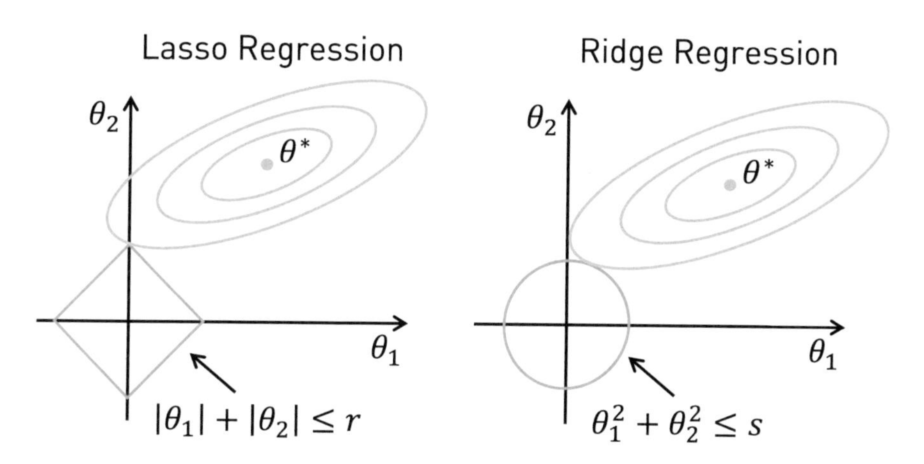

Fig. 3.2 Illustration of lasso regression vs. ridge regression

The optimization problem of lasso regression is similar to linear regression, except for the regularization part [5].

$$\begin{aligned}
\hat{y}_i &= \theta^T x_i \\
l(y_i, \hat{y}_i) &= (\hat{y}_i - y_i)^2 \\
f(\theta) &= \frac{1}{N}\sum_{i=1}^{N} l(y_i, \hat{y}_i) + \lambda||\theta||_1 \\
f'(\theta) &= \frac{1}{N}\sum_{i=1}^{N} 2(\hat{y}_i - y_i)x_i + \lambda sign(\theta)
\end{aligned} \tag{3.10}$$

where $sign(\theta_j) = 1$ if $\theta_j > 0$, $sign(\theta_j) = -1$ if $\theta_j < 0$, and $sign(\theta_j) = 0$ if $\theta_j = 0$. The update rule can be accordingly represented as

$$\theta_{j,t+1} = \theta_{j,t} - \frac{\eta}{N}\sum_{i=1}^{N} \frac{\partial l(y_i, \hat{y}_i)}{\partial \theta_{j,t}} - \eta\lambda sign(\theta_{j,t}) \tag{3.11}$$

However, this update rule does not guarantee that the fitted coefficients become exactly zero.

Carpenter et al. [6] propose an alternative approach. The update process of coefficients is divided into two steps. The first step updates the coefficients without considering the $l1$ penalty term. The second step applies the $l1$ penalty to the coefficients. The specific coefficient is clipped when it crosses zero.

$$\begin{aligned}
\theta_{j,t+\frac{1}{2}} &= \theta_{j,t} - \frac{\eta}{N}\sum_{i=1}^{N} \frac{\partial l(y_i, \hat{y}_i)}{\partial \theta_{j,t}} \\
\theta_{j,t} &= \begin{cases} max(0, \theta_{j,t+\frac{1}{2}} - \eta\lambda) & if\ \theta_{j,t+\frac{1}{2}} > 0 \\ min(0, \theta_{j,t+\frac{1}{2}} + \eta\lambda) & if\ \theta_{j,t+\frac{1}{2}} < 0 \end{cases}
\end{aligned} \tag{3.12}$$

Logistic Regression

Logistic regression is another type of generalized linear model. Despite its name, logistic regression is actually a two-class classification model. The probability distribution is assumed to be a Bernoulli distribution, and the link function is known as logistic function [7].

$$
\begin{aligned}
\eta &= x^T\theta \\
\eta &= g(\mu) = \log_e \frac{\mu}{1-\mu} \\
\mu &= g^{-1}(\eta) = \frac{1}{1+e^{-\eta}} \\
y &\sim Bernoulli(\mu)
\end{aligned}
\tag{3.13}
$$

The optimization problem of logistic regression is formalized as follows: the class probability, the prediction decision, and the objective function [8].

$$
\begin{aligned}
&P(\hat{y}=1) = h_\theta(x) = \frac{1}{1+e^{-\theta^T x}} \\
&P(\hat{y}=0) = 1 - h_\theta(x) \\
&\hat{y} = \begin{cases} 1 & if\ \ P(\hat{y}=1) > 0.5 \\ 0 & otherwise \end{cases} \\
&f(\theta) = -\frac{1}{N}\sum_{i=1}^{N}(y_i \log P(\hat{y}_i=1) + (1-y_i)\log P(\hat{y}_i=0)) + \frac{\lambda}{2}||\theta||_2
\end{aligned}
\tag{3.14}
$$

Here, $P(\hat{y} = 1)$ denotes the probability that the predicted output (label) equals 1, and $P(\hat{y} = 0)$ denotes the probability of the predicted output being 0. $P(\hat{y} = 1)$ is also known as the sigmoid function or logistic function, as illustrated in Fig. 3.3. The sigmoid function has several properties: (1) it takes a real number as input and

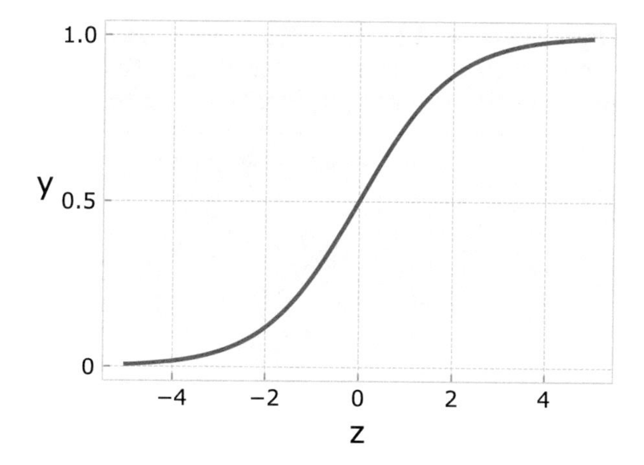

Fig. 3.3 Illustration of the sigmoid function

maps it into the range (0, 1), which can be treated as a probability, (2) it is roughly liner near $z = 0$, with a sharp slope, pushing large values of z toward 0 or 1, and (3) it is differentiable so that gradient optimization algorithms can be applied.

If the linear predictor of the coefficients θ and an input instance x_i is fed into the sigmoid function, it outputs a value between 0 and 1, which is considered as the probability of predicted output as 1. With this probability, the predicted output is decided with a threshold of 0.5—that is, $\hat{y}$ is predicted as 1 when $P(\hat{y} = 1)$ is larger than 0.5; otherwise, the predicted value is 0.

The loss function of logistic regression is defined as the cross-entropy loss which measures how close the predicted output $\hat{y}$ is to the true output y. Cross-entropy loss is a negative log-likelihood loss that does conditional maximum likelihood estimation—maximize the log probability of the correct prediction:

$$\begin{aligned} \log P(y|x) &= y \log P(\hat{y} = 1) + (1 - y) \log P(\hat{y} = 0) \\ &= y \log \frac{1}{1 + e^{-\theta^T x}} + (1 - y) \log \left(1 - \frac{1}{1 + e^{-\theta^T x}}\right) \end{aligned} \tag{3.15}$$

Equation 3.15 needs to be maximized, which is equal to minimizing the cross-entropy loss below:

$$\begin{aligned} l(x, y; \theta) = -\log P(y|x) &= -[y \log P(\hat{y} = 1) + (1 - y) \log P(\hat{y} = 0)] \\ &= -\left[y \log \frac{1}{1 + e^{-\theta^T x}} + (1 - y) \log \left(1 - \frac{1}{1 + e^{-\theta^T x}}\right)\right] \end{aligned} \tag{3.16}$$

An intuitive explanation for the adoption of cross-entropy loss is that the fitted classifier should do the best to produce probability 1 to the correct predicted output ($\hat{y} = y = 1$ or $\hat{y} = y = 0$) and probability 0 to the incorrect predicted output ($\hat{y} = 1, y = 0$ or $\hat{y} = 0, y = 1$). The range of the cross-entropy loss (negative log of probability) is from 0 (all predictions are correct, log of 1) to infinity (all predictions are incorrect, log of 0). Therefore, the fitting of coefficients in logistic regression is equal to the minimization of the cross-entropy loss, which can be solved by gradient optimization algorithms. The gradient computation and update rule of gradient optimization algorithms are given below:

$$\begin{aligned} \theta^* &= arg \min_{\theta} f(\theta) \\ f'(\theta_t) &= \frac{1}{N} \sum_{i=1}^{N} \left(\frac{1}{1 + e^{-\theta_t^T x_i}} - y_i\right) x_i + \lambda \theta_t \\ \theta_{t+1} &= \theta_t - \eta f'(\theta_t) \end{aligned} \tag{3.17}$$

Multinomial Logistic Regression

Logistic regression is a classification model for two-class datasets. It cannot handle datasets with multiple classes, e.g., multilabel image recognition, protein analysis, and weather forecasting. Multinomial logistic regression, also called softmax regression, is a simple extension of binary logistic regression that can support more than two categories of the dependent variables (a.k.a. desired outputs, response variables) [9]. In multinomial logistic regression, since the predicted outputs have more than two classes, the model should provide the probability of the predicted output $\hat{y}$ being in each potential class $P(\hat{y} = c|x)$.

Assume there are k different classes, given the input variable x_i and the coefficients vector $\boldsymbol{\theta} = [\theta_1, \theta_2, \ldots, \theta_k]$ for each class, the algorithm gives the linear predictor for each class: $\mathbf{z} = [\theta_1^T, \theta_2^T, \ldots, \theta_k^T]x$. Next, the vector z is transformed into the probabilities of all the classes, each of which is between 0 and 1 and their sum equals 1. Multinomial logistic regression adopts the softmax function, an extension of the sigmoid function to obtain these probabilities:

$$softmax(z_i) = \frac{e^{z_i}}{\sum_{j=1}^{k} e^{z_j}}, \quad 1 \le i \le k \tag{3.18}$$

The denominator $\sum_{j=1}^{k} e^{z_j}$ normalizes all the values to probabilities so that the softmax function assures that every probability is in the range (0,1). The softmax of the linear predictor vector $\mathbf{z}$ is

$$softmax(\mathbf{z}) = \left[\frac{e^{z_1}}{\sum_{j=1}^{k} e^{z_j}}, \frac{e^{z_2}}{\sum_{j=1}^{k} e^{z_j}}, \ldots, \frac{e^{z_k}}{\sum_{j=1}^{k} e^{z_j}}\right] \tag{3.19}$$

This form guarantees that all the probabilities sum up to 1. With the above softmax functions, the possibility of each class is naturally the corresponding value in $softmax(\mathbf{z})$:

$$P(\hat{y} = c|x) = \frac{e^{z_c}}{\sum_{j=1}^{k} e^{z_j}} \tag{3.20}$$

The predicted output of x can be easily obtained by choosing the class having the highest probability.

The objective function of multinomial logistic regression adopts the softmax function rather than the sigmoid function. The loss over one instance x_i is defined as the sum of the log of softmax probabilities over all the classes, a.k.a. cross-entropy loss.

$$l(x, y; \theta) = -\sum_{j=1}^{k} 1\{y = j\} \log P(\hat{y} = j|x) = -\sum_{j=1}^{k} 1\{y = j\} \log \frac{e^{\theta_j^T x}}{\sum_{s=1}^{k} e^{\theta_s^T x}} \tag{3.21}$$

where the indicator function $1\{y = k\}$ is equal to 1 if the condition is satisfied. The objective over multiple inputs with regularization is therefore defined as

$$f(x, y; \theta) = \frac{1}{N}\sum_{i=1}^{N} l(x_i, y_i; \theta) + \frac{\lambda}{2}\sum_{j=1}^{k} ||\theta_j||_2 \tag{3.22}$$

Due to the indicator function $1\{\cdot\}$, the gradient of one input instance is only taken for the true class c and the probability that the softmax function produces for class c:

$$\begin{aligned} l'(x, y; \theta_c) &= \frac{\partial l(x, y; \theta)}{\partial \theta_c} \\ &= -\frac{\partial}{\partial \theta_c}\left[1(y = c)\log\frac{e^{\theta_c^T x}}{\sum_{j=1}^{k} e^{\theta_j^T x}}\right] \\ &= -[1\{y = c\} - P(\hat{y} = c|x)]x \\ &= -\left[1\{y = c\} - \frac{e^{\theta_c^T x}}{\sum_{j=1}^{k} e^{\theta_j^T x}}\right]x \\ \frac{\partial f(x, y; \theta)}{\partial \theta_c} &= \frac{1}{N}\sum_{i=1}^{N} l'(x_i, y_i; \theta_c) + \lambda\theta_c \\ \theta_{c,t+1} &= \theta_{c,t} - \eta\frac{\partial f(x, y; \theta)}{\partial \theta_c} \end{aligned} \tag{3.23}$$

With the above gradient optimization, every coefficient θ_j in the coefficient vector $\boldsymbol{\theta}$ is updated accordingly.

Note that logistic regression is a special case of multinomial logistic regression. In other words, multinomial logistic regression generalizes logistic regression to more than two classes [8].

Support Vector Machine

Support vector machine, abbreviated as SVM, is a widely used classification model [10, 11]. The basic idea of SVM is to find a hyperplane to separate the input variables, which are represented as data points in multidimensional space. This hyperplane is properly chosen to make the separation gap as wide as possible.

Assuming that the dimension (the number of features) of input variables is d, they can be seen as data points in a d-dimensional space. The classification model needs to find a $(d-1)$-dimensional hyperplane that separates these data points as accurately as possible. The goal of SVM is to obtain a hyperplane that has the maximum margin, i.e., the maximum distance between the data points and the hyperplane for both classes. From another perspective, maximizing the separation margin assures that future data points can be classified with higher confidence.

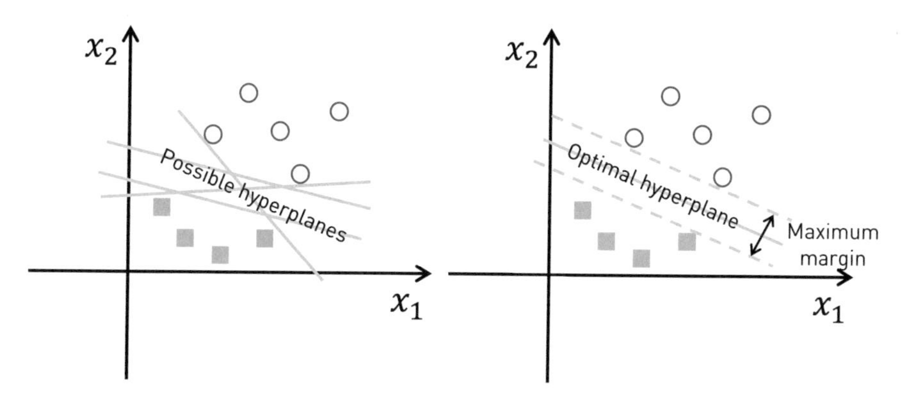

Fig. 3.4 Separation hyperplanes

Figure 3.4 shows the separating hyperplanes in SVM. There are several data points in this two-dimensional space, which belong to two categories. To classify these data points, the classification model aims at finding a decision boundary, as a form of the hyperplane, to separate two classes. The hyperplane is a line in a two-dimensional space, a two-dimensional plane in a three-dimensional space, and a $(d-1)$-dimensional hyperplane in a d-dimensional space. With this separating hyperplane, the data points on either side of the hyperplane are assigned to different classes. Apparently, there are many possible hyperplanes that can separate two categories, as the left plotting in Fig. 3.4 illustrates. However, since some of these hyperplanes are too close to the data points, a future data point might risk locating at the wrong side of the hyperplane and is classified as the wrong class.

To increase the classification confidence for future variables, SVM chooses to find a hyperplane that has the "largest margin" between two classes. Considering each class has one data point nearest to the hyperplane, the distance between it and the hyperplane can be retreated as the margin for both classes. If this margin is maximized, the hyperplane found is called the maximum margin hyperplane. The data points that are closest to the hyperplane are called the "support vectors." As the name implies, SVM uses the support vectors to determine the position of the hyperplanes.

The input of SVM is a set of variables $\{x_1, x_2, \ldots, x_N\}$ and their desired outputs $\{y_1, y_2, \ldots, y_N\} \in \{-1, +1\}$. The linear predictor combines the coefficients θ and the input to generate the predicted output $\hat{y}$ with a bias b: $\hat{y} = \theta^T x + b$. As shown in Fig. 3.5, SVM defines three hyperplanes—one hyperplane that separates the input into two groups and two hyperplanes that cross the support vectors:

$$\begin{aligned} Separating\ hyperplane\ H_0 &: \ \theta^T x + b = 0 \\ Positive\ hyperplane\ H_1 &: \ \theta^T x + b = +1 \\ Negative\ hyperplane\ H_2 &: \ \theta^T x + b = -1 \end{aligned} \tag{3.24}$$

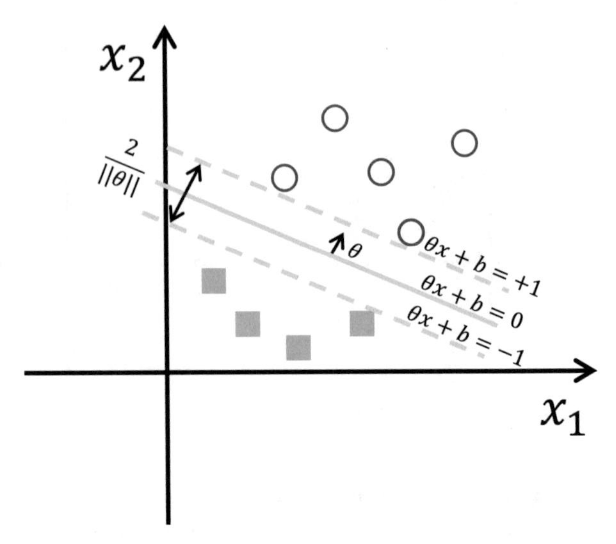

Fig. 3.5 Illustration of SVM

The data points in the area $\theta^T x + b \geq +1$ are the positive points $y = +1$ and those in the area $\theta^T x + b \leq -1$ are negative points $y = -1$. This decision can also be written as $y(\theta^T x + b) \geq 1$.

Let d_+ denote the shortest distance from the positive hyperplane to the closest positive point and d_- denote the shortest distance from the negative hyperplane to the closest negative point. The margin of the separating hyperplane is hence defined as the sum of d_+ and d_-./ Based on the definition of the positive and negative hyperplanes, the geometrical distance between these two hyperplanes is $\frac{2}{||\theta||}$ [12].

The above analysis works for data points that can be linearly separated. However, the real datasets are generally not linearly separable. In this scenario, there exist data points located at the wrong side of the separating hyperplane. SVM uses the hinge loss to introduce soft margin:

$$l(x, y; \theta) = max(0, 1 - y(\theta^T x + b)) \tag{3.25}$$

If the linear predictor $(\theta^T x + b)$ has the same sign as the desired output y and $|(\theta^T x + b)| \geq 1$, the hinge loss $l(x, t; \theta)$ is clipped to zero, meaning that accurately predicted output introduces no loss. If the signs of linear predictor and the desired output are different, the hinge loss $l(x, t; \theta) = 1 - y(\theta^T x + b) > 1$ as $y(\theta^T x + b) < 0$, meaning that hinge loss gives a positive loss for a misclassified instance. A special case is that the linear predictor and the desired output have the same sign, but the data point is between the positive hyperplane and the negative hyperplane. In this case, $y(\theta^T x + b)$ is between 0 and 1, so that hinge loss $l(x, t; \theta)$ is in the range $(0, 1)$. The more close the data point is to the separating hyperplane, the larger the hinge loss will be.

After adding the regularization term, the objective function of SVM is as follows:

$$f(x, y; \theta) = \frac{1}{N}\sum_{i=1}^{N} max(0, 1 - y_i\theta^T x_i) + \frac{\lambda}{2}||\theta||_2 \tag{3.26}$$

Here, we ignore the bias term b for simplicity. This can be achieved by adding an additional dimension to every input variable and set the value to 1. With the gradients of the hinge loss and the objective function, the coefficients can be updated accordingly [13].

$$\begin{aligned} f'(x_i, y_i; \theta) &= \frac{\partial l(x_i, y_i; \theta)}{\partial \theta} \\ &= \begin{cases} 0 & if \ \ y_i\theta^T x_i \geq 1 \\ -y\theta^T x_i & otherwise \end{cases} \\ \frac{\partial f(x, y; \theta)}{\partial \theta} &= \frac{1}{N}\sum_{i=1}^{N} f'(x_i, y_i; \theta) + \lambda\theta \\ \theta_{t+1} &= \theta_t - \eta\frac{\partial f(x, y; \theta_t)}{\partial \theta_t} \end{aligned} \tag{3.27}$$

3.1.3 *Single-Node Gradient Optimization*

In the aforementioned linear models, the coefficients can be fitted by first-order gradient optimization algorithms, including gradient descent (GD), stochastic gradient descent (SGD), and minibatch stochastic gradient descent (minibatch SGD). In this section, we describe how to execute these gradient optimization algorithms with a single machine. When they are executed in a single machine, due to the sequential nature of these gradient optimization algorithms, the simplest implementation is a serial execution with a single thread. A faster implementation with modern hardware is to execute gradient optimization algorithms with multithreading. Without loss of generalization, we will provide an overview of the state-of-the-art methods in this section, taking minibatch SGD as an example.

3.1.3.1 Serial Gradient Optimization

The serial execution of an iterative minibatch SGD algorithm is presented in Algorithm 1.

1. At each iteration, a minibatch of b training samples are taken from the dataset, denoted by $\{x_1, x_2, \ldots, x_b\}$.

Algorithm 1 Iterative minibatch SGD

---

x: features of training samples, y: desired outputs of training samples
$f(x, y; \theta)$: objective function, θ: model parameter (coefficients), θ^0: initial model parameter
T: # iterations, N: # samples, b: batch size, η: learning rate
1: $\theta^1 = \theta^0$
2: **for** $t = 1$ *to* T **do**
3: Take b training samples $\{x_1, x_2, \ldots, x_b\}$ from the dataset
4: Compute loss $l = \frac{1}{b}\sum_{i=1}^{b} f(x_i, y_i; \theta^t)$
5: Compute gradient $g = \frac{1}{b}\sum_{i=1}^{b} \frac{\partial f(x_i, y_i; \theta^t)}{\partial \theta^t}$;
6: Update the model parameter $\theta^{t+1} = \theta^t - \eta g$;
7: **end for**

---

2. Compute the loss and gradients of the objective function $f(\cdot)$ over the minibatch using the current model parameter θ^t.
3. Update the model parameter θ^t toward the opposite direction of the gradient with a learning rate η.
4. Proceed to the next iteration with the new state of the model parameter.

This training paradigm works for the linear models mentioned in this section since they can be summarized in the same form under the abstraction of the generalized linear model.

3.1.3.2 Single-Node Parallel Gradient Optimization

Since modern machines are normally equipped with multiple cores, executing gradient optimization algorithms with only one thread is a waste of resources. Many researchers choose to use multicore architecture to perform execution in parallel. Typically, multiple threads are launched to compute gradients and update the model parameters stored in the shared-memory. The main difference between the existing methods is how the training data are managed and how the shared model parameters are updated. Below, we introduce several classical single-node parallel methods, representing different training patterns, including Hogwild! [14], Delayed-SGD [15], and Dimmwitted [16].

Delayed-SGD

Langford et al. propose a multithreading parallel stochastic gradient descent with delayed updates, called Delayed-SGD [15]. They assume (1) there are W processors in a multicore machine and (2) computing the gradient of the objective function $f(\theta)$ is at least W times as expensive as it is to update the model parameter θ (read, add, and write). Delayed-SGD is a data-parallel SGD algorithm that horizontally partitions the training instances among W processors so that each processor has a training subset. The processors share one common model parameter θ, which is stored in memory.

Algorithm 2 Delayed stochastic gradient descent

x: features of training samples, y: desired outputs of training samples
$f(x, y; \theta)$: objective function, θ: model parameter (coefficients), θ^0: initial model parameter
W: # thread, τ: delay, T: # iterations, N: # samples, b: batch size, η: learning rate

1: Initialize $\theta^1, \theta^2, \ldots, \theta^\tau = 0$, compute corresponding gradient $g^t = \frac{\partial f(\theta^t)}{\partial \theta^t}$
2: **for** $t = \tau + 1$ *to* $T + \tau$ **do**
3: Take b training samples $\{x_1, x_2, \ldots, x_b\}$ from the dataset
4: Compute loss $l = \frac{1}{b} \sum_{i=1}^{b} f(x_i, y_i; \theta^t)$
5: Compute gradient $g^t = \frac{1}{b} \sum_{i=1}^{b} \frac{\partial f(x_i, y_i; \theta^t)}{\partial \theta^t}$
6: Update the model parameter $\theta^{t+1} = \theta^t - \eta g^{t-\tau}$
7: **end for**

Algorithm 2 shows the processing steps of Delayed-SGD. Each processor computes its own gradient $g^t = \frac{\partial f(\theta^t)}{\partial \theta^t}$ over a minibatch of its allocated training data. However, Delayed-SGD does not update the shared model parameter with the current gradient g^t. Instead, the processors update the shared model parameter in a round-robin fashion—Delayed-SGD allows each processor in a round-robin fashion to update the model parameter θ one at a time. Therefore, there will be a delay of $\tau = W - 1$ between the point when gradients are computed and the point when the gradients are applied to the shared model parameter. In other words, the model parameter is updated with a delayed gradient $g^{t-\tau}$ that was computed τ iterations before.

In the implementation of Delayed-SGD, the individual processors do not need to explicitly perform thread-level synchronization with each other. Delayed-SGD leverages a read/write-locking mechanism for the model parameter θ or enforces atomic updates on the model parameter. The authors also theoretically analyze the convergence rate of Delayed-SGD and prove Delayed-SGD's regret $R = \sum_{t=1}^{T} [f(\theta^t) - f(\theta^*)]$ is bounded by $R \leq 4FL\sqrt{\tau T}$. Here, the assumptions are (1) the objective function $f(\theta)$ is convex, (2) the subdifferentials are bounded $||\nabla f(\theta)|| \leq L$ by some constant $L > 0$, and (3) $f(\theta)$ is minimized at θ^*.

Referring to the basic techniques in Sect. 2, Delayed-SGD belongs to a specific point in the design space. For parallelism, Delayed-SGD belongs to the category of data parallelism as the training dataset is partitioned over the threads. For parameter sharing, Delayed-SGD uses a shared-memory architecture in which all the threads read and write the shared model parameters. For synchronization, Delayed-SGD entails a fixed delay among the threads so that it belongs to stale synchronization protocol.

Hogwild!

Although Delayed-SGD provides a convergence guarantee, it inevitably incurs inferior convergence due to delayed update and locking overhead on the model

Algorithm 3 Hogwild!

x: features of training samples, y: desired outputs of training samples
$f(x, y; \theta)$: objective function, θ: model parameter (coefficients), θ^0: initial model parameter
W: # thread, T: # iterations, N: # samples, b: batch size, η: learning rate

1: Start W threads, initialize $\theta = \theta^0$
2: **for all** $w \in \{1, \ldots, W\}$ threads **parallel do**
3: **for** $t = 1$ *to* T **do**
4: Take a training sample x_i from the dataset
5: Compute loss $l = f(x_i, y_i; \theta^t)$
6: Compute gradient $g = \frac{\partial f(x_i, y_i; \theta^t)}{\partial \theta^t}$
7: **for** $j = 1$ *to* d **do**
8: Update the j-th dimension of the model parameter $\theta_j^{t+1} = \theta_j^t - \eta g_j^t$
9: **end for**
10: **end for**
11: **end for**

parameter. Niu et al. [14] present a single-node parallel SGD algorithm without any locking. They present an update scheme called HOGWILD! which allows processors access to shared-memory with the possibility of overwriting each other's work.

Algorithm 3 presents the algorithmic details. Hogwild! assumes a shared-memory model with M processors. The model parameter θ is stored in the main memory and accessible to all the processors. Each processor can read θ and can contribute updates to θ. The component-wise addition operation on the shared model parameter is assumed to be atomic:

$$\theta_j = \theta_j + a \tag{3.28}$$

The update operation can be performed by any processor using a scalar a for θ_j, the j-th dimension of the model parameter. Hogwild! does not use a locking mechanism when multiple threads update the model parameter simultaneously. This non-locking strategy is supported by the most modern hardware, e.g., CPUs and GPUs.

At each iteration, Hogwild! samples a training instance x_i from the dataset, computes the gradient of f at x_i with the current state of the model parameter, and updates every dimension of the model parameter. The processor updates only the dimension indexed by j at a time, leaving all other dimensions alone.

Obviously, Hogwild! adopts asynchronous protocol among the processors. Therefore, the model parameter θ^t is often updated with a stale gradient, which is computed with a value of the model parameter several cycles earlier. The authors provide theoretical analysis and prove a $\frac{1}{k}$ rate of convergence for a constant learning rate scheme, both in serial and in parallel settings.

Another challenge of Hogwild! is that there could be multiple processors that write the same shared model parameter at the same time, causing conflict writes and overwrite problem. Hogwild! takes advantage of data sparsity in machine learning problems, that is, the generated gradients are sparse as the training data are sparse. For these sparse machine learning workloads, Hogwild! achieves near-linear speedups.

Regarding the fundamental techniques of the design space mentioned in Sect. 2, Hogwild! chooses data parallelism, shared-memory, and asynchronous protocol.

Dimmwitted

The above single-node parallel gradient optimization algorithms adopt certain data replication method (i.e., how to assign the dataset across threads), certain data access pattern (i.e., how the parallel threads access the assigned training data), and certain model replication method (i.e., how the parallel threads access the model parameter). None of them have studied the trade-off space regarding the data access method, model replication strategies, and data replication strategies. Zhang and Ré perform a study of the trade-off space of access methods and replication to support statistical analytics using first-order gradient optimization algorithms executed in the main memory of a Non-Uniform Memory Access (NUMA) machine [16]. Based on the anatomy, they develop a prototype engine, called Dimmwitted, and evaluate real datasets.

Two metrics are considered in this trade-off study—(1) statistical efficiency (how many steps are needed until convergence to a given tolerance) and (2) hardware efficiency (how efficiently those steps can be carried out). They identify three trade-offs that have not been explored in the literature: (1) access methods for the data, (2) model replication, and (3) data replication. Concretely, the following techniques are presented below:

- *Data Access Methods*
  - *Row-wise access.* The system scans each row of the dataset and applies a function to it and then updates the model parameter. This access method is widely used by gradient optimization algorithms.
  - *Column-wise access.* The system scans each column of the dataset and reads only the corresponding component of the model. When updating the model, the method typically writes a single component of the model. This method is often used by stochastic coordinate descent [17].
  - *Column-to-row access.* The system iterates conceptually over the columns, typically for space matrices. When iterating a column, it reads the rows in which the column is nonzero and only updates the corresponding of the model. This method is used by nonlinear support vector machines in GraphLab [18] and is the de facto approach for Gibbs sampling [19].
- *Model Replication*
  - *PerCore.* The PerCore strategy is a shared-nothing architecture, where each core (thread) maintains a mutable state of the model parameter. These core-level states are combined to form a global state of the model parameter at

a certain point (typically at the end of each epoch). PerCore is popularly used by state-of-the-art machine learning frameworks such as Spark [20] and GraphLab [18]. In the implementation of Dimmwitted, each worker (core) has its own model replica and updates the replica during the training.
- *PerMachine.* The PerMachine strategy maintains a single model replica that is shared and updated by all the workers, which is adopted by Hogwild! and Delayed-SGD.
- *PerNode.* In the architecture of an NUMA machine, there are multiple NUMA nodes and each node has multiple cores. The PerNode strategy is a hybrid of PerCore and PerMachine. Each NUMA node maintains a single model replica that is shared by all the cores in the node.

- *Data Replication*
  - *Sharding.* The most widely adopted data replication method is sharding, which partitions the dataset over all the workers and allocates one partition to each worker.
  - *Full-PerCore.* Instead of partitioning the dataset, the Full-PerCore method replicates the whole dataset for each core in the machine.
  - *Full-PerNode.* Alternatively, the Full-PerNode method replicates the whole dataset for each NUMA node. Note that each core in an NUMA node accesses the data in a different order so that the dataset in a node provides non-redundant statistical information for the cores.

The authors conduct extensive experiments to study the trade-off space of the above three techniques. Their observations and analysis are summarized below:

- *Anatomy of Data Access.* Row-wise and column-wise access methods are evaluated over several datasets. They have similar statistical efficiency—they consume a comparable number of epochs to converge to a given threshold of error or loss. However, they show significantly different hardware efficiencies. The performance is largely affected by the sparsity of the dataset, which is represented by a defined "cost ratio": $(1+\alpha)\sum_i \frac{n_i}{\sum_i n_i^2+\alpha d}$, where n_i is the number of nonzero elements of the i-th row of the dataset and α is the cost ratio between writing and reading. They find a crossover point w.r.t. the hardware efficiency (time used per epoch): when the cost ratio is small, row-wise access outperforms column-wise access as the column-wise method reads more data; on the other hand, when the ratio is large, the column-wise access outperforms the row-wise access because the column-wise method has lower write contention.
- *Anatomy of Model Replication.* The authors compare the PerCore, PerNode, and PerMachine methods for model replication. In terms of statistical efficiency, they observe that PerMachine often takes fewer epochs to converge to the same loss compared to PerNode, and PerNode uses fewer epochs than PerCore. For hardware efficiency, PerNode uses much less time to execute an epoch than PerMachine.

Algorithm 4 MR-BSP-SGD

x: features of training samples, y: desired outputs of training samples
$f(x, y; \theta)$: objective function, θ: model parameter (coefficients), θ^0: initial model parameter
W: # distributed workers, T: # iterations, R: # local update rounds, N: # samples, b: batch size,
η: learning rate

1: Horizontally partition the training dataset over W workers
2: Each worker initializes the local model parameter as θ^0
3: **for** $t = 1$ *to* T **do**
4:     **for all** $w \in \{1, \dots, W\}$ workers **parallel do**
5:         Take a minibatch of training samples $\{x_1, x_2, \dots, x_b\}$ from the dataset
6:         Compute loss $l = \sum_{i=1}^{b} f(x_i, y_i; \theta^t)$
7:         Compute gradient $g_w^t = \sum_{i=1}^{b} \frac{\partial f(x_i, y_i; \theta^t)}{\partial \theta^t}$;
8:         Aggregate the gradients from all the workers $g^t = \frac{1}{bW} \sum_{w=1}^{W} g_w^t$
9:         Update the local model parameter $\theta^{t+1} = \theta^t - \eta g^t$
10:     **end for**
11: **end for**

- *Anatomy of Data Replication.* The empirical study of data replication compares Sharding and Full-PerNode. The Full-PerCore replication method is not considered because the Full-PerNode approach dominates the Full-PerCore approach, as reads from the same node go to the same NUMA memory. In terms of statistical efficiency, the Full-PerNode method uses fewer epochs, especially for a tolerance of low error. The reason is that each model replica sees more data than Sharding and therefore produces a better estimate. Nevertheless, the Full-PerNode method is slower than the Sharding method since each epoch of Full-PerNode processes more data than Sharding.

Through this in-depth trade-off study, the authors implement Dimmwitted considering different design choices. Dimmwitted achieves significant speedups compared with other competitor systems that only use specific choices.

3.1.4 Distributed Gradient Optimization

The previous section introduces gradient optimization algorithms for linear models on a single node. As the trend of increasing available data, distributed gradient optimization for machine learning has attracted much interest. In this section, we give an overview of distributed gradient optimization algorithms by introducing several representative approaches.

3.1.4.1 MR-BSP-SGD

Chu et al. [21] design a general parallel method for many machine learning algorithms with the MapReduce framework. Algorithm 4 shows their implementation for linear models, e.g., linear regression and logistic regression. The main steps are listed below:

1. The training dataset is partitioned into W partitions, each of which is assigned to one worker.
2. Each worker initializes the local model parameters with the same value and performs T iterations of minibatch SGD.
3. At the t-th iteration, each worker takes a minibatch of training samples from the local data partition and computes the gradient of the loss function over the minibatch.
4. The local gradients of all the workers are aggregated through a mapper operator and a reducer operator.
5. Each worker updates the local model parameters with the aggregated gradients.

Except for linear models, the authors also apply this parallel scheme on other models such as K-Means, Neural Network, and Principal Components Analysis (PCA). In summary, MR-BSP-SGD chooses the design space of data parallelism, shared-nothing architecture (MapReduce) for parameter sharing, and bulk synchronous protocol, as listed in Table 3.2.

3.1.4.2 MR-MA-SGD

Zinkevich et al. [22] propose a distributed SGD algorithm under MapReduce framework, which we call MR-MA-SGD. The algorithmic logics are presented in Algorithm 5.

1. The training dataset is horizontally partitioned over W distributed workers. Each worker has a local replica of model parameters that is initialized with the same value.
2. Each worker performs T rounds of stochastic gradient descent. Each round of SGD takes a training sample x_i from the local data partition and computes the loss and gradient of the model parameter θ^t over x_i and updates the local model parameters with the gradients.
3. A master routine is used to aggregate the local model parameters and calculate their average as the output model parameter.

The computation pattern of MR-MA-SGD is model averaging that solves subproblems on each worker and finally averages local solutions to obtain a joint solution. Model average is also applied in other works [23, 24]. Different from the previously described MR-BSP-SGD, which averages local gradients at the end of each iteration, MR-MA-SGD lets each worker run SGD and finally takes the average for the local model parameters. The authors prove the convergence guarantee for MR-MA-SGD under assumptions of Lipschitz continuity, Holder continuity, Lipschitz seminorm, Hölder seminorm, and contraction. The design space of MR-MA-SGD consists of data parallelism, shared-nothing architecture (MapReduce), and bulk synchronous protocol.

Table 3.2 Summary of distributed gradient optimization algorithms

Algorithm	Parallelism	Parameter sharing	Synchronization protocol	Transferred statistics
MR-BSP-SGD	Data-parallel	MapReduce	Synchronous	Gradient
MR-MA-SGD	Data-parallel	MapReduce	Synchronous	Model
PS-BSP-SGD	Data-parallel	Parameter server	Synchronous	Gradient/update
PS-SSP-SGD	Data-parallel	Parameter server	Stale synchronous	Gradient/update
Column-SGD	Data-parallel (vertical)	MapReduce	Synchronous	Scalar (dot product)
PS-ASP-SGD	Data-parallel	Parameter server	Asynchronous	Gradient
Decentralized-PSGD	Data-parallel	Shared-nothing (MPI)	Synchronous	Model
Decentralized-ASP-SGD	Data-parallel	Shared-nothing (MPI)	Asynchronous	Model
QSGD	Data-parallel	Shared-nothing (MPI)	Synchronous	Quantized gradient
GradDrop-SGD	Data-parallel	Parameter server	Synchronous	Sparsified gradient
AlexNet	model-parallel	Parameter server	Synchronous	Gradient
SINGA	data/model-parallel	Parameter server	Synchronous	Gradient
GBDT-Spark	Data-parallel	MapReduce	Synchronous	Gradient histogram
XGBoost	Data-parallel	Shared-nothing (MPI)	Synchronous	Gradient histogram
DimBoost	Data-parallel	Parameter server	Synchronous	Gradient histogram
LightGBM	Data-parallel	Reduce-scatter	Synchronous	Gradient histogram
Vero	Data-parallel (vertical)	MapReduce	Synchronous	Scalar (splits)

3.1.4.3 PS-BSP-SGD

MR-BSP-SGD and MR-MA-SGD both implement distributed gradient optimization algorithms over the MapReduce architecture. However, MapReduce-based distributed machine learning relies on a single node to be responsible for aggregating the statistics, either gradients or model parameters. This causes a serious system bottleneck when the transferred data are large or there are many machines.

To address the system bottleneck of MapReduce-based systems, some researchers propose the parameter server architecture and implement distributed gradient optimization algorithms accordingly. Figure 3.6 is the architecture

Algorithm 5 MR-MA-SGD

x: features of training samples, y: desired outputs of training samples
$f(x, y; \theta)$: objective function, θ: model parameter (coefficients), θ^0: initial model parameter
W: # distributed workers, T: # iterations, R: # local update rounds, N: # samples, η: learning rate

1: **for all** $w \in \{1, \ldots, W\}$ workers **parallel do**
2: Initialize the model parameter $\theta_w^1 = \theta^0$
3: **for** $t = 1$ *to* T **do**
4: Take a training sample x_i from the dataset
5: Compute loss $l = f(x_i, y_i; \theta^t)$
6: Compute gradient $g_w^t = \frac{\partial f(x_i, y_i; \theta^t)}{\partial \theta^t}$;
7: Update the local model parameter $\theta_w^{t+1} = \theta^t - \eta g_w^t$;
8: **end for**
9: **end for**
10: Aggregate local model parameters θ_w^{t+1} from all workers
11: Compute the global model parameter $\theta^{t+1} = \frac{1}{W}\sum_{w=1}^{W} \theta_w^{t+1}$
12: Send the global model parameter θ^{t+1} to all the workers.

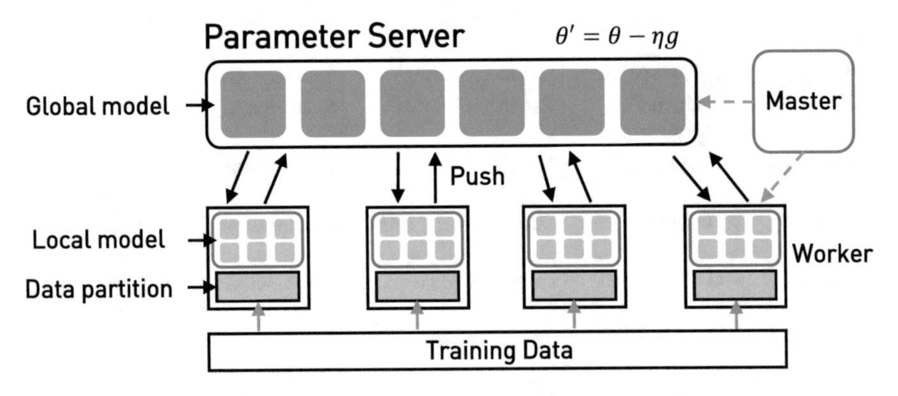

Fig. 3.6 Parameter server architecture

of parameter server. There are three types of nodes in the parameter server architecture:

- *Parameter server.* Several parameter server nodes together store a global copy of the model parameter θ. The model parameter, typically represented as a vector or a matrix, is partitioned over these servers, and each server node stores a partition. The parameter server provides two interfaces: (1) *Pull*: worker nodes can obtain the current model parameter via the pull interface and (2) *Push*: worker nodes can use the push interface to send data to the parameter server.
- *Worker.* Each worker node maintains a local copy of the model parameter and executes gradient computation using its assigned training data. Periodically, each worker pulls the current global model parameter from the servers, computes intermediate statistics (e.g., gradients, model updates), and pushes the statistics to the parameter servers.

Algorithm 6 PS-BSP-SGD

x: features of training samples, y: desired outputs of training samples
$f(x, y; \theta)$: objective function, θ: model parameter (coefficients), θ^0: initial model parameter
W: # distributed workers, P: # parameter servers, T: # iterations, N: # samples, η: learning rate

Worker $w = 1, \ldots, W$:

1: **for** $t = 1$ *to* T **do**
2: $\theta^t_w = PS.Pull(w, t)$
3: Take a minibatch of training samples $\{x_1, x_2, \ldots, x_b\}$ from the dataset
4: Compute loss $l = \sum_{i=1}^{b} f(x_i, y_i; \theta^t_w)$
5: Compute gradient $g^t_w = \sum_{i=1}^{b} \frac{\partial f(x_i, y_i; \theta^t_w)}{\partial \theta^t_w}$;
6: $PS.Push(g^t_w, t)$
7: **end for**

Parameter Server $p = 1, \ldots, P$:

1: Initialize $\theta = \theta^0$
2: **function** $Push\left(g^t_w, t\right)$:
3: $\theta = \theta - \eta g^t_w$
4:
5: **function** $Pull\,(w, t)$:
6: **if** all the workers finish the $(t-1)$-th iteration **then**
7: **return** θ
8: **end if**

- *Master.* The platform launches a master node that is responsible for node management, synchronization, and failure recovery.

A series of works have designed synchronous SGD algorithms with the parameter server architecture, such as Parameter Server [25], Petuum [26], and Angel [27]. We call this training scheme PS-BSP-SGD and present the processing procedure below:

1. The training dataset is partitioned over the workers. Each worker reads its assigned partition from the data source and maintains a local copy of the modal parameter.
2. At the t-th iteration, each worker pulls the current model parameter from the parameter server via the *Pull* interface. Then, the worker computes a gradient over a minibatch of training instances and pushes the gradient to the parameter server via the *Push* interface.
3. On the parameter server side, each server maintains a partition of the model parameter. When receiving a push request from a worker, the server directly adds the gradient to the stored model partition. Once received a pull request from a worker, the server first checks whether all the workers have finished the previous iteration—if so, the server sends the current model partition to the worker; otherwise, the server waits until the condition is satisfied.

In the above algorithm, the intermediate statistics that are transmitted are gradients. However, it is not the case for all the scenarios. Some works choose to update the local model replica with gradients and transfer the update of the model to

Algorithm 7 PS-SSP-SGD

x: features of training samples, y: desired outputs of training samples
$f(x, y; \theta)$: objective function, θ: model parameter (coefficients), θ^0: initial model parameter
W: # distributed workers, P: # parameter servers, T: # iterations, N: # samples, η: learning rate

s: staleness threshold, t_{min}: iteration of the slowest worker

Worker $w = 1, \ldots, W$:

1: **for** $t = 1$ *to* T **do**
2: $\theta_w^t = PS.Pull(w, t)$
3: Take a minibatch of training samples $\{x_1, x_2, \ldots, x_b\}$ from the dataset
4: Compute loss $l = \sum_{i=1}^{b} f(x_i, y_i; \theta_w^t)$
5: Compute gradient $g_w^t = \sum_{i=1}^{b} \frac{\partial f(x_i, y_i; \theta_w^t)}{\partial \theta_w^t}$;
6: $PS.Push(g_w^t, t)$
7: **end for**

Parameter Server $p = 1, \ldots, P$:

1: Initialize $\theta = \theta^0$, $t_{min} = 0$
2: **function** $Push\left(g_w^t, t\right)$:
3: $\theta = \theta - \eta g_w^t$
4: **if** all the workers finish the t-th iteration **then**
5: **return** $t_{min} = t$
6: **end if**
7:
8: **function** $Pull\,(w, t)$:
9: **if** $t \leq t_{min} + s$ **then**
10: **return** θ
11: **end if**

the parameter server [28]. In general, the design space of PS-BSP-SGD is composed of data parallelism, shared-memory architecture, and synchronous protocol.

3.1.4.4 PS-SSP-SGD

PS-BSP-SGD relies on the strict bulk synchronous protocol to synchronize the workers, which might suffer the straggler problem in heterogeneous environments. A solution to this heterogeneity problem is to relax the synchronization. Ho et al. [29] propose a stale synchronous parallel protocol, SSP for short, and implement SSP in a parameter server system. Instead of setting a strict barrier at the end of each iteration (or epoch), SSP entails a more flexible rule that the faster worker can run at most s iterations than the slower workers, where s is a user-defined staleness threshold.

Algorithm 7 elaborates the details of the SGD algorithm under SSP synchronization protocol, called PS-SSP-SGD. On the worker side, the processing procedure is the same as PS-BSP-SGD. Each worker pulls the latest model parameter from the parameter server. Nevertheless, the processing on the parameter server side is more complicated than that of PS-BSP-SGD. To control the speed gap between different workers, the server maintains a variable t_{min} that records the current iteration of the slowest worker. In the *push* function, when the parameter server receives a gradient g_w^t (or update) from one worker that has finished the t-th iteration, it updates the global model parameter with a learning rate. If all the workers have finished the t-th iteration, the parameter server updates the minimum iteration t_{min} to t. This can be easily implemented by recording the number of pushed gradients at each iteration. In the *pull* function, once a request is received in which the worker provides the current finished iteration t, the parameter server first checks the SSP condition, i.e., t does not exceed the minimum iteration t_{min} more than s iterations. The authors prove the correct convergence of SSP and give the bound of regret:

$$R[\mathbf{X}] = \left[\frac{1}{T}\sum_{t=1}^{T} f(\theta^t)\right] - f(\theta^*) \leq 4FL\sqrt{\frac{2(s+1)W}{T}} \tag{3.29}$$

where θ^* is the optimal solution of the model parameter. F and L are certain constants of the two assumptions below.

Assumption 1 (L-Lipschitz Function) For a convex function $f(\theta)$, the subdifferentials are bounded by some constant L: $\|\nabla f(\theta)\| \leq L$.

Assumption 2 (Bounded Diameter) For any $\theta, \theta' \in \mathbb{R}^n$ and some constant $F > 0$, $D(\theta\|\theta') = \frac{1}{2}\|\theta - \theta'\|^2 \leq F^2$ holds.

The design space of PS-SSP-SGD is obvious—data parallelism, shared-memory architecture, and stale synchronization protocol. The statistics transferred between the workers and the servers are gradient or model updates.

3.1.4.5 Column-SGD

From a data management perspective, the above methods all partition the training dataset by rows (a.k.a. horizontal partitioning). Using this row-oriented partitioning strategy, each worker communicates a complete copy of the whole model with other workers (or the parameter servers). When the trained model is large, row-oriented partitioning faces expensive communication costs.

An alternative to row partitioning is column partitioning, which vertically partitions the training dataset so that each worker is responsible for some columns (or features). The model parameters are also partitioned over the workers using the same layout as the partition of training data. Since each worker only maintains a portion of the model parameters, the communication overhead between the workers can be largely decreased.

Zhang et al. [30] propose a column-partitioning scheme called ColumnSGD for generalized linear models. In ColumngSGD, the training dataset and the model parameters are partitioned by columns, leading to a layout where corresponding data

Algorithm 8 ColumnSGD

x: features of training samples, y: desired outputs of training samples
$f(x, y; \theta)$: objective function, θ: model parameter (coefficients), θ^0: initial model parameter
W: # distributed workers, T: # iterations, N: # samples, b: batch size, η: learning rate

1: Use the same layout to vertically partition the training dataset and the model parameters over W workers
2: Each worker initializes the local model parameter as θ^0
3: **for** $t = 1$ *to* T **do**
4:     **for all** $w \in \{1, \ldots, W\}$ workers **parallel do**
5:         Take a minibatch of training samples X_B from the local vertical partition
6:         Compute statistics using X_B and the corresponding local partition of model
7:         Aggregate the statistics from all the workers via a master node
8:         Compute the gradients using the sum of statistics and X_B
9:         Update the local model parameter using the gradients
10:     **end for**
11: **end for**

and model partitions are collocated on the same machine. The authors summarize that the gradient computation of linear models involves scalar statistics. Taking logistic regression as an example, the gradient g over a minibatch of data X_B is

$$g(\theta, X_B) = \sum_{x_i \in X_B} \frac{-y_i}{1 + \exp(y_i \cdot \sum_{j=1}^{d}(\theta_j \cdot x_{ij}))} \tag{3.30}$$

where x_i and y_i are the features and labels of the i-th training instance, θ is the model parameter, d is the dimension of the feature, and x_{ij} denotes the j-th feature of x_i. As can be seen, $\sum_{j=1}^{d}(\theta_j \cdot x_{ij})$ is the sum of the dot products over the columns of the feature and the model parameter. This sum operator can be decomposed into several partial sums, each of which is taken over a subset of the columns. ColumnSGD leverages this property by partitioning the training data vertically and letting each worker handle a partial sum in the gradient computation.

The procedure of ColumnSGD is given in Algorithm 8. In ColumnSGD, there are several worker nodes and one master node. The training dataset and the model parameters are vertically partitioned using the same partitioning scheme. Each worker is assigned one vertical partition of the training dataset and one vertical partition of the model parameters. In this way, the data and model partitions are "collocated" on each worker. Each iteration of the training includes the following steps:

1. Each worker randomly samples a minibatch of training instances from its responsible vertical data partition and computes "statistics" using the minibatch and the local vertical model partition. For instance, statistics in training LR are in the form of dot products.
2. The master aggregates the statistics from all the workers and broadcasts the aggregated statistics to the workers. The aggregation function is usually the sum of the statistics.

3. Each worker receives aggregated statistics from the master and computes the gradient.
4. The local vertical model partition is updated using the gradients.

The strategy of column partitioning is also adopted in training other types of distributed machine learning models. Ordentlich et al. [31] explore column partitioning for reducing the communication cost in the word2vec model [32]. Coordinate descent (CD) is an optimization technique that naturally accesses the training data in a column-wise manner. ColumnML is proposed to train coordinate descent over column-partitioned in-database training data [33].

3.1.4.6 Other Related works

Other than the above methods, there are other distributed gradient optimization algorithms that can train linear models. A-BSP-SGD (Arbitrarily sized Bulk Synchronous Parallel SGD) [34] is proposed to extend PS-BSP-SGD by allowing a strict synchronization every several iterations instead of one iteration. ESSP-SGD (Eager SSP SGD) [35] is proposed as a variant of SSP-SGD in which the server eagerly propagates the new state of model parameters to workers so that staleness between workers is reduced. MLlib* [36] implements model averaging and parameter server-style communication in Spark to accelerate the communication.

3.2 Neural Network Models

We next introduce distributed gradient optimization algorithms designed for deep learning models. Note that most of the previously described methods can be applied to neural network models. Nevertheless, as the blooming development of deep learning, many researchers propose methods specially designed for training deep learning models in a distributed setting.

3.2.1 Formalization of Neural Network

In this section, we provide the basics of feed-forward neural networks for classification problems.[1] In the terminology of neural network, the input is $\mathbf{X} = \sum_{i=1}^{N} x_i$ called features (or variables), and the output is $\mathbf{Y} = \sum_{i=1}^{N} y_i$ called labels (or response variables). Each pair (x_i, y_i) is called a training instance. The goal of a neural network model is to find a statistical model $f(\theta)$, also called a classifier,

[1] Neural networks can be easily extended to regression problems.

which is parametrized by the model parameter θ. For each class $k \in \{1, 1, \ldots, K\}$, the classifier gives the probability of the predicted output $P(\hat{y}_i = k|x_i)$. Finally, the class with the highest probability is chosen as the predicted class. For example, the multinomial logistic regression introduced in Sect. 3.1 can be used in neural network models to produce probabilities with nonlinear transformation.

3.2.1.1 Model Definition

Generally, neural network models can be formalized as the stacking of nonlinear functions, as illustrated in Fig. 3.7. The structure of the model contains an input layer, several hidden layers, and an output layer. Each layer contains several nodes and each node has an activation function σ:

$$h^{(l)} = g^{(l)}(h^{(l-1)}) = \sigma(W^{(l)}h^{l-1)} + b^{(l)}) \tag{3.31}$$

where $h^{(l)}$ denotes the output of the l-th layer ($l = \{1, \ldots, L\}$, especially $h^{(0)} = \mathbf{x}$). $W^{(l)}$ denotes the weight matrix and $b^{(l)}$ denotes the bias at the l-th layer. $\sigma(\cdot)$ is a nonlinear activation function. At the l-th layer, the activation function takes $h^{(l-1)}$ as the input and generates an output $h^{(l)}$. Popular activation functions include sigmoid function, tanh function, and ReLU function [37].

The objective function can be defined by the output of the last layer $h^{(L)}$ and the desired label y. Since the outputs are multilabels, the loss function is normally the multinomial logistic loss. The softmax function is used to compute the probabilities using the linear predictors:

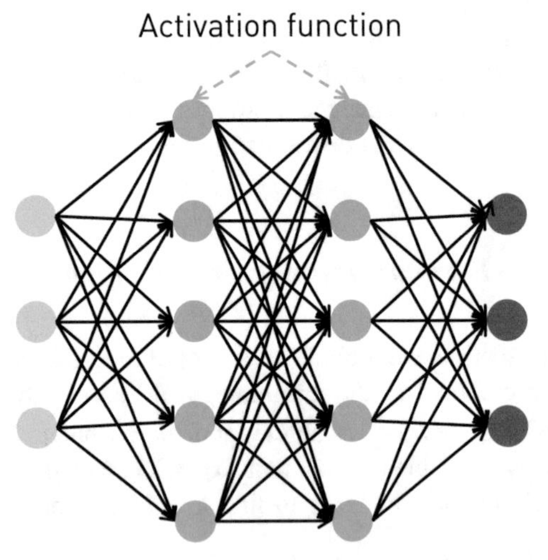

Fig. 3.7 Structure of a feed-forward neural network

$$\mathbf{z} = W^{(L+1)}h^{(L)} + b^{(L+1)} \in \mathbb{R}^K$$
$$p_k = softmax(z_k) = \frac{\exp^{z_k}}{\sum_k \exp^{z_k}}, k \in \{1, .., K\} \quad (3.32)$$

Finally, the loss function is defined as the cross-entropy loss using the outputs of the softmax function:

$$l(x, y; \boldsymbol{\theta}) = -\sum_{k=1}^{K} 1\{y = k\} \log p_k \quad (3.33)$$

The weight matrices and biases are summarized as the model parameter $\boldsymbol{\theta} = \{W^{(l)}, b^{(l)}; l \in \{1, \ldots, K\}\}$. For a sequence of training instances, the objective function $f(\boldsymbol{\theta})$ is defined as the sum of the loss function over all the instances and the $l2$-norm regularization of the model parameter $\boldsymbol{\theta}$.

3.2.1.2 Back-Propagation

To find the optimal model parameter $\boldsymbol{\theta}$, the objective function needs to be minimized. Gradient optimization algorithms, such as stochastic gradient descent and minibatch stochastic gradient descent, are often chosen to perform the minimization. However, different from linear models where the model parameter can be directly updated once the gradients are obtained, neural networks have multiple layers which make the gradient computation and model update more complex. Back-propagation is leveraged to compute the gradients in neural network models [38]. Fundamentally, back-propagation introduces a chain rule across connected layers. The gradient computation starts at the last layer of the neural network, and the gradients are "back-propagated" toward the first layer—$\{\frac{\partial f}{\partial h^{(L)}}, \frac{\partial f}{\partial h^{(L-1)}}, \ldots, \frac{\partial f}{\partial h^{(1)}}\}$ are computed successively. Concretely, the gradient at the l-th layer is computed by

$$\frac{\partial f}{\partial h^{(l-1)}} = \frac{\partial h^{(l)}}{\partial h^{(l-1)}} \frac{\partial f}{\partial h^{(l)}} = \frac{\partial \sigma(W^{(l)}h^{(l-1)} + b^{(l)})}{\partial h^{(l-1)}} \frac{\partial f}{\partial h^{(l)}} \quad (3.34)$$

These derivatives are then used to update the weight matrices:

$$W^{(l)} = W^{(l-1)} - \eta \frac{\partial f}{\partial W^{(l)}}, \text{ where } \frac{\partial f}{\partial W^{(l)}_{ij}} = \frac{\partial f}{\partial h_i^{(l)}} \sigma' h_j^{(l)} \quad (3.35)$$

where σ' is the gradient of the activation function. For example, if the activation function is ReLU, $\sigma' = 1$ if the j-th element of $(W^{(l)}h^{l-1)} + b^{(l)})$ is not negative; otherwise, $\sigma' = 0$.

For a minibatch of input instances, the entire computation process contains two phases—the forward phase and the backward phase. In the forward phase, the model feeds the input to the first layer and calculates the output of each layer. In the backward phase, the model uses the chain rule to calculate the gradient of each layer and updates the model parameters accordingly.

3.2.2 Overview of Popular Neural Network Models

This section offers a brief overview of the popular state-of-the-art neural network models, including AutoEncoder, Deep Belief Network, Convolutional Neural Network, Recurrent Neural Network, Generative Adversarial Network, and Deep Reinforcement Learning. Since this book does not focus on the area of neural networks, the readers can refer to [39–41] for more details.

3.2.2.1 AutoEncoder

AutoEncoder is an unsupervised machine learning model that aims at reducing the dimensionality of input data and preserving important information. Like other dimension reduction methods such as PCA (Principal Component Analysis), the goal of AutoEncoder is to learn an encoder function $f : \mathbb{R}^d \rightarrow \mathbb{R}^k (k < d)$ to transform the input data $\mathbf{x} \in \mathbb{R}_d$ to a low-dimensional space $h \in \mathbb{R}^k$ (also called representation) and a decoder function $g : \mathbb{R}^k \rightarrow \mathbb{R}^d$ such that the difference between $\mathbf{x}$ and $g(f(\mathbf{x}))$ is minimized. In AutoEncoder, f and g are both neural network models, and the objective function is defined as (Fig. 3.8)

$$min_{f,g} \frac{1}{N} \sum_{i=1}^{N} \|x_i - g(f(x_i))\|_2^2 \tag{3.36}$$

AutoEncoder methods can be categorized into Sparse AutoEncoder and Denoising AutoEncoder. Sparse AutoEncoder uses techniques such as $l1$-norm regularization to introduce sparsity into the model [42]. Since only a small number of the hidden nodes are activated and others are inactivated, the learned representations are sparse. Denoising AutoEncoder is designed to increase model robustness for noisy input data or input data with missing values [43]. Typically, Denoising AutoEncoder introduces random noise to the original data and trains a model over the built new input data.

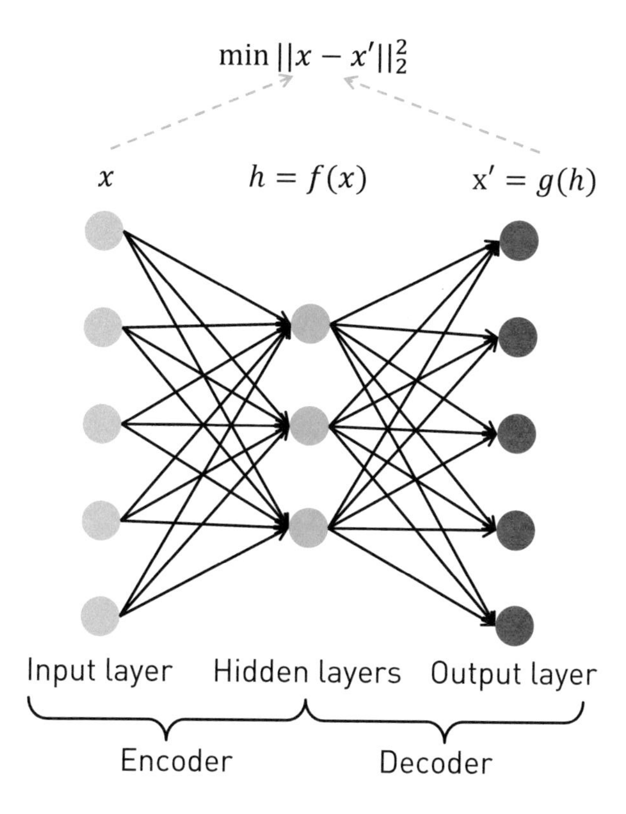

Fig. 3.8 Structure of an AutoEncoder model

3.2.2.2 Deep Belief Network

Deep belief network (DBN) [44–46] is a kind of deep learning model by stacking several restricted Boltzmann machines (RBMs) [47]. Figure 3.9 shows the structure of RBM and DBN. RBM has one visible layer and one hidden layer. Although derived from the two-layer structure of the Boltzmann machine, restricted Boltzmann machine has no connection between nodes in the same layer and only allows connection between the visible layer and the hidden layer. Since DBN consists of RBMs, there are connections between the layers but not between nodes within one layer. Each RBM's hidden layer serves as the visible layer for the next RBM module. DBN can be trained in an unsupervised layer-wise fashion, that is, it trains one layer at a time and adds more layers successively. Then, the whole network is fine-tuned using back-propagation.

3.2.2.3 Convolutional Neural Network

Convolutional neural network (CNN), mainly designed for image classification tasks, is a series of neural network models that uses techniques of the local receptive

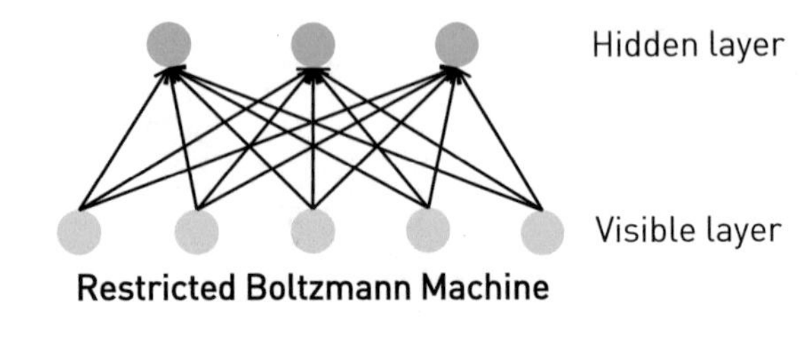

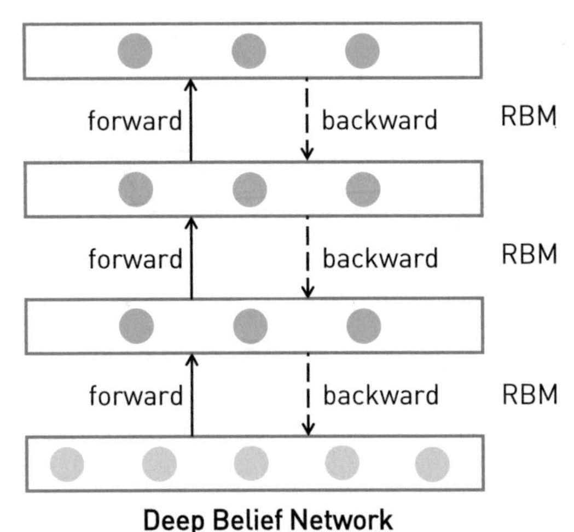

Fig. 3.9 Structure of restricted Boltzmann machine and deep belief network

field and weight sharing. Typically, a CNN model consists of several convolutional layers, pooling layers, and fully connected layers.

- *Convolutional layer.* A convolutional layer uses several filters to extract local features from the raw images or the previous layers. After passing through a convolutional layer, the input image becomes a feature map. Each filter is represented as a tensor $F \in \mathbb{R}^{d_1 \times d_1 \times d_2}$, where d_1 is the filter size and d_2 is the number of filters. Each filter can be seen as a local receptive field for the input neurons of each layer. Since the weights of the filters are shared for all the neurons in a layer, the model size of a convolutional layer is small, allowing the network to be deeper.
- *Pooling layer.* A pooling layer is used to reduce the dimensionality of feature maps and keep the invariance of the network by aggregating neighboring features to a single value. For example, a $d \times d$ max-pooling layer chooses the maximum value in a $d \times d$ area in the input of the layer. Commonly used pooling layers include max-pooling, average-pooling, and global-pooling.
- *Fully connected layer.* A fully connected layer connects each neuron in one layer to every neuron in the next layer. A fully connected layer is often used as the last layer in a CNN model that outputs classification results.

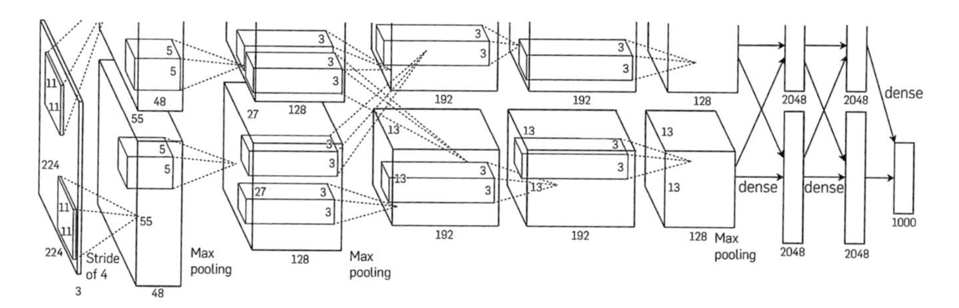

Fig. 3.10 Structure of AlexNet

In a neural network, the neurons have different receptive fields because they receive inputs from different neurons in the previous layer. The input area of a neuron is called its receptive field. In a convolutional layer, each neuron receives input from a small area of the previous layer according to the filters; therefore, the receptive area is much smaller than the entire previous layer. While in a fully connected layer, each neuron receives input from every neuron of the previous layer; the receptive field is thereby the entire previous layer.

In recent decades, convolution neural networks have drawn much attention from researchers. Many CNN models have been proposed, including LeNet [48], AlexNet [49], ResNet [50], VGGNet [51], and GoogleNet [52].

Taking AlexNet as an example, Fig. 3.10 shows its network structure. Overall, AlexNet consists of five convolutional layers and three fully connected layers. The output of the last convolutional layer is fed to a 1000-way multinomial logistic regression. These five convolutional layers use different filters, as specified in the figure.

3.2.2.4 Recurrent Neural Network

Recurrent neural networks (RNNs) are another type of neural network that handles sequence data or time-series data. Different from convolutional neural networks, RNNs allow previous outputs to be used as the inputs for the current state, while traditional neural networks assume that the inputs and outputs are independent of each other.

The structure of the vanilla RNN model is illustrated in Fig. 3.11. For a given time-series input sequence $\{x_1, x_2, \ldots, x_T\}$, the hidden state h_t and the output y_t at timestamp t are represented as

$$\begin{aligned} h_t &= f(W_{hh}h_{t-1} + W_{xh}x_t + b_h) \\ y_t &= \sigma(W_{hy}h_t + b_y) \end{aligned} \tag{3.37}$$

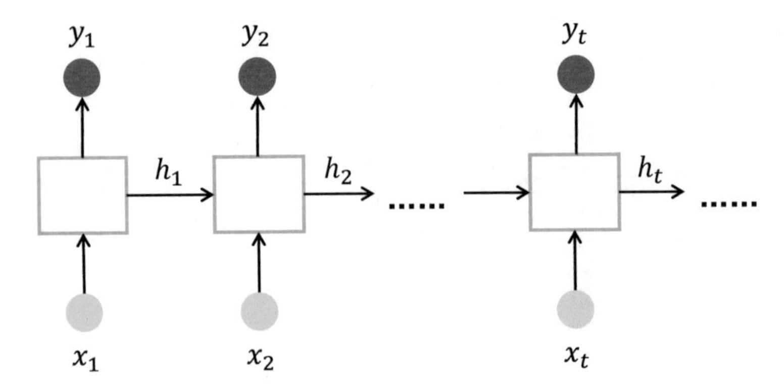

Fig. 3.11 Structure of a vanilla RNN model

where f and σ are nonlinear activation functions (such as sigmoid, tanh, and ReLU). W_{hh}, W_{xh}, and W_{hy} are the weight matrices, and b_h and b_y are the biases. These model parameters are generally trained by back-propagation.

However, the vanilla RNN model suffers from the vanishing gradient problem during back-propagation. As a result, vanilla RNN can only reserve short-term memory and cannot capture long-term dependencies in the input sequence when the length is long. To resolve this problem, researchers design advanced RNN models such as GRUs (gated recurrent units) [53] and LSTM (long short-term memory) [54, 55]. GRUs propose a gate structure that has the same length as the hidden state h_t. The values in the gate vector are between 0 and 1, which are used to perform elementwise multiplication with the previous hidden state h_{t-1}. Intuitively, the gate structure determines how much the previous hidden states are reserved. LSTM also uses three gate structures to control the values kept across layers—the forget gate, the input gate, and the output gate. Besides, LSTM introduces a memory of cell state to record sequence information (e.g., singular/plural form in a sentence) and therefore reserves long-term dependencies.

3.2.2.5 Other Neural Networks

In addition to the above introduced neural networks, there are some advanced types of neural networks, e.g., generative adversarial network, and deep reinforcement learning. Generative adversarial network (GAN) [56] is an unsupervised neural network model in which two neural networks contest with each other in the form of a zero-sum game, where one agent's gain is another agent's loss. Typically, GAN trains a generator and a discriminator. The generator generates synthesized data given randomized input based on a predefined distribution. The discriminator evaluates the synthesized data from the generator. The goal of GAN is to fool the discriminator by letting the generator produce synthesized data that are classified as not synthesized by the discriminator. Deep reinforcement learning (DRL) combines

Algorithm 9 PS-ASP-SGD

x: features of training samples, y: desired outputs of training samples
$f(x, y; \theta)$: objective function, θ: model parameter (coefficients), θ^0: initial model parameter
W: # distributed workers, P: # parameter servers, T: # iterations, N: # samples, η: learning rate

Worker $w = 1, \ldots, W$:

1: Initialize the local model replica as θ^0
2: **for** $t = 1$ *to* T **do**
3: $\theta_w^t = PS.Pull(w, t)$
4: Take a minibatch of training samples $\{x_1, x_2, \ldots, x_b\}$ from the dataset
5: Compute loss $l = \sum_{i=1}^{b} f(x_i, y_i; \theta_w^t)$
6: Compute gradient $g_w^t = \sum_{i=1}^{b} \frac{\partial f(x_i, y_i; \theta_w^t)}{\partial \theta_w^t}$;
7: $PS.Push(g_w^t, t)$
8: **end for**

Parameter Server $p = 1, \ldots, P$:

1: Initialize $\theta = \theta^0$
2: **function** $Push\left(g_w^t, t\right)$:
3: $\theta = \theta - \eta g_w^t$
4:
5: **function** $Pull(w, t)$:
6: **return** θ

the techniques of reinforcement learning and deep learning [57]. DRL trains neural networks to make decisions through trial and error. Due to the space constraint and the target of this book, we do not describe each type of these models in detail.

3.2.3 *Distributed Gradient Optimization*

3.2.3.1 PS-ASP-SGD

Researchers in Google propose a distributed machine learning framework, named DistBelief, which implements a parameter server architecture to distributedly store model parameters [28]. Based on DistBelief, they design Downpour SGD (also called PS-ASP-SGD in this work), an asynchronous distributed stochastic gradient descent algorithm.

The methods of data management and model management are the same as PS-BSP-SGD. The training dataset is partitioned over the workers, and each worker is allocated a partition. The model parameters are also partitioned over the parameter servers, and each parameter server holds a model shard. As has been shown in Algorithm 9, the parameter server provides two interfaces for the worker—the *Pull* function and the *Push* function. Each worker uses the *Pull* function to obtain the latest model parameters from the parameter serve and uses the *Push* function to send the statistics to the parameter server. The difference between PS-ASP-SGD and PS-BSP-SGD is that PS-ASP-SGD chooses the asynchronous parallel protocol to synchronize the workers.

The procedure of the workers is the same as PS-BSP-SGD. Each worker maintains a local copy of the model parameters. At the t-th iteration, each worker (1) pulls the current state of the model from the parameter servers, (2) computes a gradient using a minibatch of training data, (3) pushes the gradient to the parameter servers, and (4) starts the next iteration.

Under the asynchronous protocol, the workers do not have to wait for each other, and each worker can process the next iteration once the current iteration is finished. The asynchronous protocol is realized on the parameter servers. On the one hand, the *push* function directly applies the received statistics to the stored model parameter shard. On the other hand, the *pull* function directly sends back the stored model parameter shard to the requesting worker. Since the parameter server incurs no locking or waiting mechanism, the workers can run under the ASP protocol.

PS-ASP-SGD reveals its advantages in several aspects. First, PS-ASP-SGD achieves higher hardware efficiency (i.e., less time for one iteration) without any waiting overhead. Second, when there are some stragglers in a heterogeneous environment, PS-ASP-SGD avoids the system being stalled by the stragglers. Third, PS-ASP-SGD is more robust to system failures than its BSP-based counterpart. If one machine fails, a BSP-based SGD algorithm has to entirely stop until the failure is recovered, while an ASP-based SGD algorithm can continue training with the remaining model shards. However, since the ASP protocol allows an arbitrary speed gap between the workers, PS-ASP-SGD inevitably introduces a staleness problem that one worker might read stale model parameters and therefore compute stale gradients. The authors propose integrating the AdaGrad learning rate scheduler which adaptively adjusts the learning rate according to the historical change of the gradients. An empirical study shows that this strategy can address the stability concern of PS-ASP-SGD.

3.2.3.2 Decentralized-PSGD

Gradient optimization algorithms designed over MapReduce or parameter server essentially belong to the category of centralized training. They rely on a single node or several nodes to aggregate the statistics. However, the performance of centralized training will be significantly degraded when the network bandwidth is low. This motivates researchers to study algorithms for decentralized topologies, where all nodes can only communicate with their neighbors, and there is no such a central node, as shown in Fig. 3.12.

Lian et al. [58] study a decentralized parallel stochastic gradient descent algorithm on the decentralized network, called decentralized-PSGD (D-PSGD for short). Their theoretical analysis indicates that D-PSGD admits similar total computational complexity but requires much less communication for the busiest node. The algorithm details are illustrated in Algorithm 10. The decentralized communication topology is defined by an undirected weighted graph (V, M). V

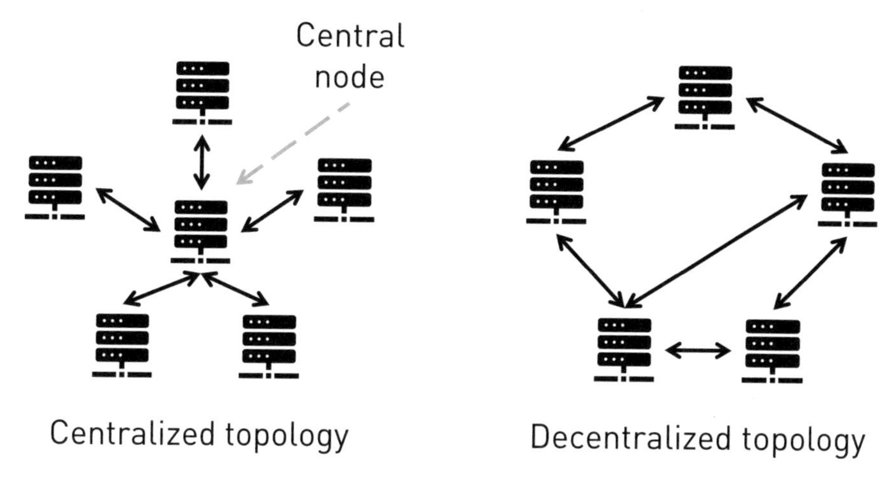

Fig. 3.12 An illustration of centralized topology and decentralized topology

---

Algorithm 10 Decentralized-PSGD

---

x: features of training samples, y: desired outputs of training samples
$f(x, y; \theta)$: objective function, θ: model parameter (coefficients), θ^0: initial model parameter
W: # distributed workers, P: # parameter servers, T: # iterations, N: # samples, η: learning rate,
M: weight matrix

Worker $w = 1, \ldots, W$:

1: Initialize $\theta_w^1 = \theta_0$
2: **for** $t = 1$ *to* T **do**
3: Take a minibatch of training samples $\{x_1, x_2, \ldots, x_b\}$ from the dataset
4: Compute loss $l = \sum_{i=1}^{b} f(x_i, y_i; \theta_w^t)$
5: Compute gradient $g_w^t = \sum_{i=1}^{b} \frac{\partial f(x_i, y_i; \theta_w^t)}{\partial \theta_w^t}$;
6: Fetch model parameters from necessary neighbors and compute their weighted average according to the weight matrix: $\theta_w^{t+1/2} = \sum_{j=1}^{W} W_{wj} \theta_w^t$
7: Update the local model parameter $\theta_w^{t+1} = \theta_w^{t+1/2} - \eta g_w^t$
8: **end for**

---

denotes the computational workers, and $M \in \mathbb{R}^{W \times W}$ is a symmetric stochastic weight matrix. Specifically, $M_{i,j} \in [0, 1]$, $M_{ij} = M_{ji}$, and $\sum_j M_{ij} = 1$ for all i. The usage of M is to decide the connectivity between two workers, e.g., $M_{ij} = 0$ means worker i and worker j are totally disconnected.

Each worker maintains a local copy of the model parameter and runs the logics in Algorithm 10. Each iteration contains the following steps:

1. Each worker samples one minibatch of training instances and computes the gradient using the batch.
2. Each worker exchanges its local model parameter θ_w^t with its neighbors according to the weight matrix M and averages the received model parameters and its own local model parameter, generating the averaged parameter $\theta_w^{t+1/2}$.

3. Each worker updates its local model parameter using the averaged parameter and the computed local gradient.

The authors prove that D-PSGD has a linear speedup and the same convergence rate as centralized parallel SGD. The decentralized topology is chosen as the ring network topology with corresponding weight matrix M in the form of

$$M = \begin{bmatrix} \frac{1}{3} & \frac{1}{3} & 0 & \cdots\cdots & \frac{1}{3} \\ \frac{1}{3} & \frac{1}{3} & \frac{1}{3} & 0 \; \cdots & 0 \\ & & & & \\ \cdots & \cdots & \cdots & \cdots & \cdots \\ \frac{1}{3} & 0 & \cdots & 0 \; \frac{1}{3} & \frac{1}{3} \end{bmatrix} \tag{3.38}$$

meaning that the workers are organized as a ring and each worker only communicates with its two neighboring workers. The design space of D-PSGD chooses data parallelism, shared-nothing architecture, and bulk synchronous protocol. Similar decentralized gradient optimization algorithms to D-PSGD can be referred to [59–63].

3.2.3.3 Decentralized-ASP-SGD

Lian et al. further extend BSP-based decentralized SGD to an asynchronous setting [64]. When there are stragglers in the training cluster, which is common for hundreds of devices, the asynchronous training paradigm is more robust in terms of hardware efficiency. However, the implementations of most asynchronous SGD algorithms choose a centralized design, as illustrated in Fig. 3.12—a central parameter server holds the model parameters shared for all the workers. Each worker calculates gradients with its own training data and updates the shared model asynchronously. In this system architecture, the parameter server may become a communication bottleneck and slow down the overall convergence.

Motivated by the idea of decentralized-PSGD, the authors conduct a study of combining asynchronous SGD and decentralized SGD and propose an asynchronous decentralized parallel stochastic gradient descent, which we categorized as decentralized-ASP-SGD (DA-SGD). DA-SGD inherits the merits of both asynchronous training and decentralized training—the workers do not wait for all other workers and communicate in a decentralized fashion.

Algorithm 11 describes the DA-SGD algorithm. Each worker w maintains a local copy of the model parameters θ_w and repeats the following steps at each iteration:

1. *Sample data.* Randomly sample a minibatch of training data $\{x_1, x_2, \ldots, x_b\}$.
2. *Compute gradients.* Use the minibatch data to compute the stochastic gradient $g_w = \sum_{i=1}^{b} \frac{\partial f(x_i, y_i; \hat{\theta}_w)}{\partial \hat{\theta}_w}$, where $\hat{\theta}_w$ is read from the model parameters in the local memory.

Algorithm 11 Decentralized-ASP-SGD

x: features of training samples, y: desired outputs of training samples
$f(x, y; \theta)$: objective function, θ: model parameter (coefficients), θ^0: initial model parameter
W: # distributed workers, P: # parameter servers, T: # iterations, N: # samples, η: learning rate,
M: weight matrix

Worker $w = 1, \ldots, W$:

1: Initialize the local model parameter $\theta_w^1 = \theta^0$
2: **for** $t = 1$ *to* T **do**
3: Randomly sample a minibatch of training samples $\{x_1, x_2, \ldots, x_b\}$ from the dataset
4: Compute loss $l = \sum_{i=1}^{b} f(x_i, y_i; \hat{\theta}_w^t)$
5: Compute gradient $g_w^t = \sum_{i=1}^{b} \frac{\partial f(x_i, y_i; \hat{\theta}_w^t)}{\partial \hat{\theta}_w^t}$;
6: Randomly sample an averaging matrix M_w
7: Average local model parameters by $[\theta_1^{t+1/2}, \theta_2^{t+1/2}, \ldots, \theta_W^{t+1/2}] = [\theta_1^t, \theta_2^t, \ldots, \theta_W^t]M$
8: Update the local model parameter $\theta_w^{t+1} = \theta_w^{t+1/2} - \eta g_w^t$; $\theta_j^{t+1} = \theta_j^{t+1/2}, \forall j \neq w$
9: **end for**

3. *Model update.* Update the model in the local memory by $\theta_w = \theta_w - g_w$. Note that $\hat{\theta}_w$ may be different from θ_w as it can be modified by other workers in the *model averaging* step.
4. *Model averaging.* Randomly select a neighbor (worker w') and average the local model parameters with the local model parameter $\theta_{w'}$ on the worker w': $\theta_i, \theta_{i'} = \frac{\theta_w}{2} + \frac{\theta_{w'}}{2}$.

Each worker runs the above procedure independently without any global synchronization. Hence, it reduces the idleness of each worker and the overall training speed is still fast even if some workers are slow. Especially, the *model averaging* step can be further generalized into the following form for all workers:

$$[\theta_1, \theta_2, \ldots, \theta_W] = [\theta_1, \theta_2, \ldots, \theta_W]M$$

where M is a floating-point weight matrix. The authors prove that the proposed distributed SGD algorithm achieves linear speedup w.r.t. the number of workers and the same convergence bound as its synchronous and/or centralized counterparts.

3.2.3.4 QSGD

Compared with linear models, neural network models usually have larger model parameters. Consequently, neural networks generally cause much more communication costs when trained in a distributed environment. To address this system barrier, many researchers propose to compress the transferred gradients between the workers. Among these methods, a family of quantization-based methods is widely adopted by transforming floating-point gradients to a discrete set of values with limited bits. Although these methods belong to the category of lossy compression and inevitably introduce errors after compression, they have shown superior empirical performance with convergence guarantee.

Alistarh et al. propose a distributed SGD algorithm using quantization-based compression, called QSGD [65]. QSGD contains two major steps: (1) quantize each element in the gradients by random rounding and (2) an efficient lossless code for quantized gradients.

- *Stochastic Quantization.* The quantization function $Q_s(\cdot)$ converts the input value to s quantization levels. Given the hyperparameter s, there are s different uniformly distributed levels between 0 and 1, and each element in the gradient is assigned to one level. For a gradient $\mathbf{g} \in \mathbb{R}^d$, $Q(\mathbf{g})$ is defined as

$$Q_s(g_i) = \|\mathbf{g}\|_2 \cdot sign(g_i) \cdot \xi_i(\mathbf{g}, s) \tag{3.39}$$

where $sign(\cdot)$ denotes the sign of the element in the gradient, and $\xi_i(\cdot)$ denotes the quantization level defined below. Assume $0 \leq l < s$ is an integer and $\frac{g_i}{\|g\|_2} \in [\frac{l}{s}, \frac{l+1}{s}]$, then $[\frac{l}{s}, \frac{l+1}{s}]$ is the quantization interval of $\frac{g_i}{\|g\|_2}$. The quantization level of the i-th element of $\mathbf{g}$, denoted by $\xi_i(\mathbf{g}, s)$, is defined by introducing a random probability:

$$\xi_i(\mathbf{g}, s) = \begin{cases} \frac{l}{s} & with\ probability\ 1 - p(\frac{g_i}{\|g\|}, s) \\ \frac{l+1}{s} & otherwise \end{cases} \tag{3.40}$$

where $p(a, s) = as - l$ for any $a \in [0, 1]$. The distribution of $\xi_i(\mathbf{g}, s)$ has minimal variance over distributions with support $\{0, \frac{1}{s}, \ldots, 1\}$ and assures that the expectation $\mathbb{E}[\xi_i(\mathbf{g}, s)] = \frac{|g_i|}{\|\mathbf{g}\|_2}$. Specially, if $\mathbf{g} = \mathbf{0}$, $Q(\mathbf{g}, s) = \mathbf{0}$.
- *Efficient Coding of Gradients.* After stochastic quantization, the quantized output of a gradient vector $\mathbf{g}$ can be represented as a triple—($\|\mathbf{g}\|_2, \boldsymbol{\sigma}, \boldsymbol{\zeta}$). Here, vector $\boldsymbol{\sigma}$ stores the signs of the elements in $\mathbf{g}$, and vector $\boldsymbol{\zeta}$ stores the integer values $s \cdot \xi_i(\mathbf{g}, s)$. The frequencies of the integer values in $\boldsymbol{\zeta}$ are different, e.g., larger integers are less frequent. Considering this characteristic, QSGD uses an efficient coding based on specialized Elias integer coding [66]. Specifically, for a positive integer k, its Elias code $Elias(k)$ starts from the binary representation of k to which it depends on the length of this representation. It then recursively encodes this prefix. For any positive integer k, the length of the code is $|Elias(k)| = \log k + \log\log k + \ldots + 1 \leq (1 + o(1)) \log k + 1$. The advantage of this coding scheme is that encoding and decoding can be done efficiently.

The encoding and encoding of QSGD for a given gradient $\mathbf{g}$ are as follows:

- *Encoding.* QSGD uses stochastic quantization to quantize the gradient $\mathbf{g}$ and produces the quantized triple ($\|\mathbf{g}\|_2, \boldsymbol{\sigma}, \boldsymbol{\zeta}$). Then, 32 bits are used to encode $\|g\|_2$ and conduct Elias encoding for $\boldsymbol{\zeta}$. Finally, QSGD appends a bit for each σ_i.
- *Decoding.* The decoding phase first reads 32 bits to reconstruct $\|g\|_2$ and then iteratively uses the decoding method of Elias to recover the values of $\boldsymbol{\zeta}$ and σ.

Algorithm 12 QSGD

x: features of training samples, y: desired outputs of training samples
$f(x, y; \theta)$: objective function, θ: model parameter (coefficients), θ^0: initial model parameter
W: # distributed workers, T: # iterations, b: batch size, η: learning rate

Worker $w = 1, \ldots, W$:

1: Initialize the local model parameter $\theta_w^1 = \theta_0$
2: **for** $t = 1$ *to* T **do**
3: Randomly sample a minibatch of training samples $\{x_1, x_2, \ldots, x_b\}$ from the dataset
4: Compute loss $l = \sum_{i=1}^{b} f(x_i, y_i; \theta_w^t)$
5: Compute gradient $g_w^t = \sum_{i=1}^{b} \frac{\partial f(x_i, y_i; \theta_w^t)}{\partial \theta_w^t}$
6: Encode gradient $M_w = Encode(g_w^t)$
7: Broadcast quantized gradient M_w^t to all other workers
8: Receive quantized gradients $\{M_j^t\}_{j=1}^W$ from all other workers
9: Decode gradients $\{\hat{g}_j^t = Decode(M_j^t)\}_{j=1}^W$
10: Update the local model parameter $\theta_w^{t+1} = \theta_w^t - \frac{\eta}{W} \sum_{j=1}^{W} \hat{g}_j^t$
11: **end for**

Algorithm 12 shows the procedure of QSGD. Essentially, QSGD adopts data parallelism and bulk synchronous protocol for distributed training. There are W workers which communicate with each other via point-to-point messages. Each worker maintains a local copy of the model parameters θ. At the t-th iteration, the processing of each worker has the following steps:

1. Each worker computes a gradient g_w^t using a minibatch of training data.
2. The gradient is encoded into a quantized gradient M_w^t. Then, the quantized worker is broadcast to all other workers.
3. Each worker receives quantized gradients from all other workers, denoted by $\{M_j^t\}_{j=1}^W$.
4. The received quantized gradients are recovered using the decoding function, denoted as $\{\hat{g}_j^t = Decode(M_j^t)\}_{j=1}^W$.
5. The local model parameters are updated using the sum of decoded gradients.

Some works also propose quantization-based distributed gradient optimization algorithms. 1-Bit SGD [67] is proposed to quantize the gradients in only one bit and accumulate quantization errors, which can be considered a special case of QSGD. ECQ-SGD [68] improves over 1-bit SGD by accumulating all the previous quantization errors, rather than only from the last iteration as in 1-bit SGD. Jiang et al. [69] find that the uniform quantization of QSGD does not fit gradients that are generally nonuniform. They design a novel quantization-based compression method, which uses data sketches [70–73] to generate nonuniform quantization levels.

Algorithm 13 GradDrop-SGD

x: features of training samples, y: desired outputs of training samples
$f(x, y; \theta)$: objective function, θ: model parameter (coefficients), θ^0: initial model parameter
W: # distributed workers, T: # iterations, b: batch size, η: learning rate, r: dropping rate, res: gradient residuals

Worker $w = 1, \ldots, W$:

1: Initialize the local model parameter $\theta_w^1 = \theta^0$, and the local gradient residual $res = \mathbf{0}$
2: **function** $GradDrop\ (g, r)$:
3: $g = g + res$
4: Select threshold $thres$ so that $r\%$ elements of $|g|$ is smaller
5: $d = \mathbf{0}$
6: $d[i] = g[i], \forall g[i] \in g\ \ and\ \ |g[i]| > thres$
7: $res = g - d$
8: **return** $sparse(d)$
9:
10: **for** $t = 1\ \ to\ \ T$ **do**
11: $\theta_w^t = PS.Pull\ (w, t)$
12: Randomly sample a minibatch of training samples $\{x_1, x_2, \ldots, x_b\}$ from the dataset
13: Compute loss $l = \sum_{i=1}^{b} f(x_i, y_i; \theta_w^t)$
14: Compute gradient $g_w^t = \sum_{i=1}^{b} \frac{\partial f(x_i, y_i; \theta_w^t)}{\partial \theta_w^t}$
15: Encode sparsified gradient $\hat{g}_w^t = GradDrop(g_w^t, r)$
16: $PS.Push(\hat{g}_w^t, t)$
17: **end for**

Parameter Server $p = 1, \ldots, P$:

1: Initialize $\theta = \theta^0$
2: **function** $Push\ (g_w^t, t)$:
3: $\theta = \theta - \eta \cdot dense(g_w^t)$
4:
5: **function** $Pull\ (w, t)$:
6: **if** all workers finishes the $(t-1)$-th iteration **then**
7: **return** θ
8: **end if**

3.2.3.5 Sparsification-SGD

Different from quantization-based distributed gradient optimization algorithms, which compress the gradients through quantization techniques, another compression routine is to transform the gradient into a sparse form, and thereby the size of transferred data is reduced.

Aji et al. [74] state that gradients are mostly skewed as most elements are near zero. They propose a sparsification-based compression method called gradient dropping that maps 99% small elements (by absolute value) to zero before transmitting. The functionality of gradient dropping is given in Algorithm 13, called *GradDrop*. The input of the *GradDrop* is a gradient vector $\mathbf{g}$ and a dropping rate r, and a gradient residual res is maintained to accumulate the compression errors. The function (1) adds the residual to the gradient, (2) selects threshold $thres$ so that $r\%$ elements of $|g|$ is smaller than the threshold, (3) constructs a sparsified gradient d

by filtering the elements larger than the threshold, (4) computes the latest gradient residual by subtracting g and dm, and (5) returns the sparsified gradient d in the sparse format.

The authors implement a distributed SGD algorithm, which we call GradDrop-SGD, based on the proposed compression method *GradDrop* with a parameter server architecture under the bulk synchronous protocol. The details of GradDrop-SGD are presented in Algorithm 13.

1. *Compute gradient.* At the t-th iteration, each worker pulls the latest model parameters from the parameter server and computes a gradient using a minibatch of training data.
2. *Compress gradient.* The computed gradient is compressed via the *GradDrop* function.
3. *Send gradient.* The compressed gradient is sent to the parameter server. The parameter server parses the sparse gradient and updates the global model parameter accordingly.

The sparsification-based compression strategy is also adopted by other works [75, 76]. Stich et al. [77] analyze k-sparsification methods for compression (e.g., top-k or random-k) and prove that sparsification-based SGD converges at the same rate as vanilla SGD when equipped with error compensation. Lin et al. [78] state that 99.9% of the gradient exchange in distributed SGD are redundant and propose deep gradient compression (DGC) to reduce communication. As quantization-based compression causes an error, DGC engineers four methods to preserve accuracy during compression—momentum correction, local gradient clipping, momentum factor masking, and warm-up training. Alistarh et al. [79] investigate the convergence performance of sparsification-based compression methods. They prove that, under certain assumptions, data-parallel SGD via sparsifying gradients by magnitude with local error correction provides convergence guarantees for both convex and nonconvex smooth objectives. Wangni et al. [80] propose randomly dropping out each coordinate in the gradient by a probability p_i. To ensure that the resulting sparsified gradient vector $Q(g)$ is unbiased, the nonzero coordinates are amplified, from g_i to $\frac{gi}{pi}$.

3.2.3.6 Model-Parallel SGD

The above distributed gradient optimization algorithms all use a data-parallel mechanism, in which the whole training dataset is partitioned over the workers and each worker has a local copy of the entire model. Another type of parallelism is model parallelism that partitions the model over the workers.

Krizhevsky et al. [49] design a model-parallel strategy to train their proposed AlexNet model. As shown in Fig. 3.10, the whole neural network structure is partitioned into two parts, and each part is put onto one GPU. Two GPUs can access the whole training dataset and communicate with each other during the training process.

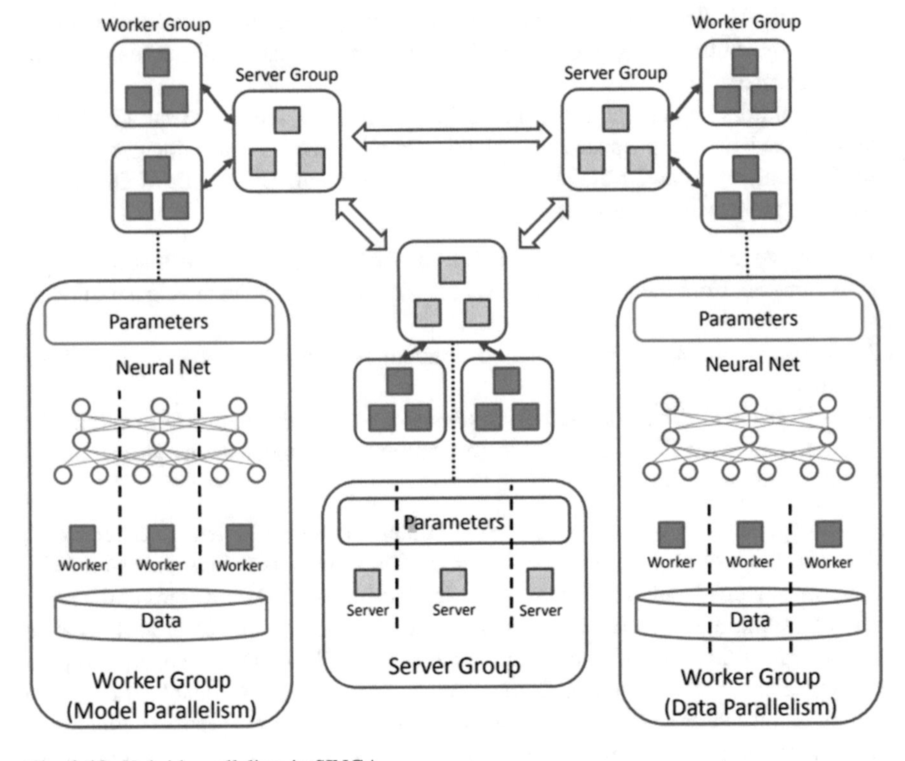

Fig. 3.13 Hybrid parallelism in SINGA

Wang et al. [81] design a deep learning framework SINGA and propose a hybrid parallel architecture, as shown in Fig. 3.13. The architecture consists of multiple server groups and worker groups. One server group has several servers, and one worker group has several workers. Each worker group communicates with one specific server group. The servers in one server group together store a complete replica of the model parameters and handle requests from their responsible worker groups. To guarantee the consistency of model replicas, different server groups synchronize their model replicas periodically. Inside one server group, each server maintains a partition of the model parameters. Each worker group trains the model replica using a partition of the training dataset and sends computed gradients to the corresponding server group. In SINGA, there are two strategies to distribute the training workload among the workers within a worker group. The first strategy is model parallelism—each worker can train a subset of parameters using all data partitioned to the worker group. The second strategy is data parallelism—each worker can train all the parameters using a subset of the data. As can be seen, SINGA supports a hybrid parallelism that combines data parallelism and model parallelism.

3.3 Gradient Boosting Decision Tree

In addition to linear models and neural network models, another type of machine learning model trained by a gradient-based optimization algorithm is an ensemble tree model called gradient boosting decision tree (GBDT). GBDT has been widely used in both academia and industrial [82]. It is also one of the most popular choices in data analytics competitions such as Kaggle and KDDCup.

3.3.1 Formalization of Gradient Boosting Decision Tree

Gradient boosting decision tree (GBDT) is a popular *tree ensemble model* [83, 84]. Figure 3.14 illustrates a GBDT model—in each tree, each training instance x_i is classified to one leaf node that predicts the instance with a weight ω. GBDT uses the regression tree that predicts a continuous weight on one leaf, rather than the decision tree in which each leaf predicts a class. GBDT sums the predictions of all the trees and gets the prediction for a single instance:

$$\hat{y}_i = \sum_{t=1}^{T} \eta f_t(x_i) \tag{3.41}$$

where T denotes the number of trees, and $f_t(x_i)$ denotes the prediction of the t-th tree.

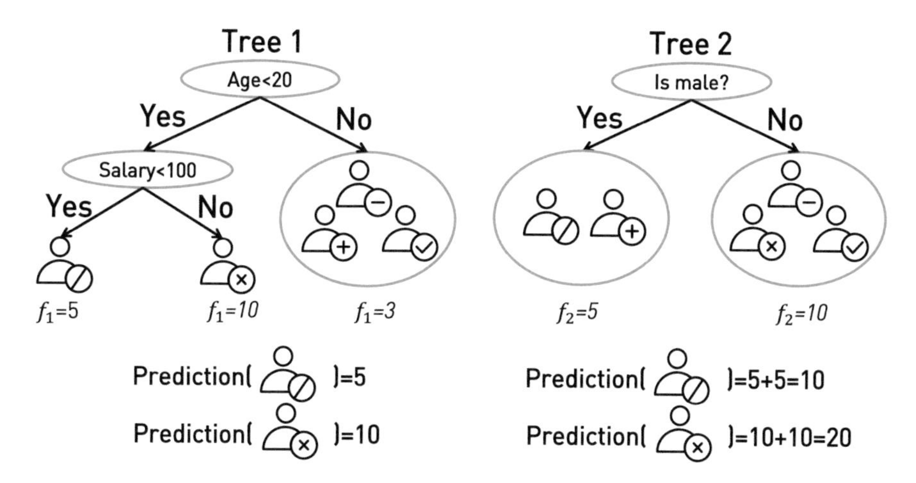

Fig. 3.14 An example of gradient boosting decision tree

Training Methodology

Unlike linear and neural models that can be trained with gradient-based optimization methods [22, 85], GBDT is trained in an additive manner. For the t-th tree, we minimize the following regularized objective:

$$F^{(t)} = \sum_{i=1}^{N} l(y_i, \hat{y}_i^{(t)}) + \Omega(f_t) = \sum_{i=1}^{N} l(y_i, \hat{y}_i^{(t-1)} + f_t(x_i)) + \Omega(f_t) \tag{3.42}$$

where l is a loss function that calculates the loss given the prediction and the target (e.g., logistic loss $l = \log(1 + e^{-y_i \hat{y}_i})$ or square loss $l = (y_i - \hat{y}_i)^2$). Ω is a regularization function for avoiding overfitting. In this work, we follow XGBoost [86] and choose Ω of the following form:

$$\Omega(f_t) = \gamma L + \frac{1}{2}\lambda\|\omega\|^2 \tag{3.43}$$

where L is the number of leaves in the tree and ω is the weight vector. γ and λ are hyperparameters. LogitBoost [87] expands $F^{(t)}$ via a second-order approximation.

$$F^{(t)} \approx \sum_{i=1}^{N} [l(y_i, \hat{y}_i^{(t-1)}) + g_i f_t(x_i) + \frac{1}{2} h_i f_t^2(x_i)] + \Omega(f_t) \tag{3.44}$$

where g_i and h_i are the first- and second-order gradients: $g_i = \nabla_{\hat{y}_i} l(y_i, \hat{y}_i)$ and $h_i = \nabla^2_{\hat{y}_i} l(y_i, \hat{y}_i)$. Let $I_j = \{i | x_i \in leaf_j\}$ be the set of instances belonging to the j-th leaf and removing the constant term, and we obtain the approximation of $F^{(t)}$:

$$\widetilde{F}^{(t)} = \sum_{j=1}^{L} \left[\left(\sum_{i \in I_j} g_i \right) \omega_j + \frac{1}{2} \left(\sum_{i \in I_j} h_i + \lambda \right) \omega_j^2 \right] + \gamma L \tag{3.45}$$

The optimal weight and objective of the j-th leaf are

$$\omega_j^* = -\frac{\sum_{i \in I_j} g_i}{\sum_{i \in I_j} h_i + \lambda}, \quad \widetilde{F}^* = -\frac{1}{2} \sum_{j=1}^{L} \frac{\left(\sum_{i \in I_j} g_i\right)^2}{\sum_{i \in I_j} h_i + \lambda} + \gamma L \tag{3.46}$$

We can enumerate every possible tree structure to find the optimal solution. However, this scheme is arguably impractical in practice. Therefore, existing research works generally adopt a greedy algorithm to successively split tree nodes, as illustrated in Algorithm 14. Given K split candidates for each feature at a given node, we enumerate all the instances belonging to this node to build a gradient histogram that summarizes the gradient statistics (lines 4–8). Specially, G_{mk} sums the first-order gradients of the instances whose m-th features fall into the range of the

Algorithm 14 Greedy splitting algorithm

M: # features, N: # instances, K: # split candidates
g_i, h_i: first-order and second-order gradients of an instance

1: **for** $m = 1$ to M **do**
2: generate K split candidates $S_m = \{s_{m1}, s_{m2}, \ldots, s_{mk}\}$
3: **end for**
4: **for** $m = 1$ to M **do**
5: loop N instances to generate gradient histogram with K buckets
6: $G_{mk} = \sum g_i$ where $s_{mk-1} < x_{im} < s_{mk}$
7: $H_{mk} = \sum h_i$ where $s_{mk-1} < x_{im} < s_{mk}$
8: **end for**
9: $gain_{max} = 0, G = \sum_{i=1}^{N} g_i, H = \sum_{i=1}^{N} h_i$
10: **for** $m = 1$ to M **do**
11: $G_L = 0, H_L = 0$
12: **for** $k = 1$ to K **do**
13: $G_L = G_L + G_{mk}, H_L = H_L + H_{mk}$
14: $G_R = G - G_L, H_R = H - H_L$
15: $gain_{max} = max\left(gain_{max}, \frac{G_L^2}{H_L+\lambda} + \frac{G_R^2}{H_R+\lambda} - \frac{G^2}{H+\lambda}\right)$
16: **end for**
17: **end for**
18: Output the split with max gain

$(k-1)$-th and the k-th split candidates, and H_{mk} sums the second-order gradients in the same manner. After building the gradient histograms for all the features, we use the histograms to find the split that gives the maximal objective gain (lines 10–17) defined as

$$Gain = \frac{1}{2}\left[\frac{\left(\sum_{i\in I_L} g_i\right)^2}{\sum_{i\in I_L} h_i + \lambda} + \frac{\left(\sum_{i\in I_R} g_i\right)^2}{\sum_{i\in I_R} h_i + \lambda} - \frac{\left(\sum_{i\in I} g_i\right)^2}{\sum_{i\in I} h_i + \lambda}\right] - \gamma \tag{3.47}$$

where I_L and I_R are the left and right child nodes after splitting. Note that we can propose split candidates with several methods. The exact method sorts all the instances by each feature and uses all possible splits. When the exact method is too time-consuming, the previous work uses percentiles of feature distribution [86, 88]. Many works have designed distributed quantile sketch algorithms for this purpose, such as GK [70], DataSketches [72], and WQS [86].

Two prevailing techniques used to prevent overfitting are shrinkage and feature sampling. Shrinkage multiplies the leaf weight in Eq. 3.41 by a hyperparameter η called the learning rate before adding it to the prediction. The feature sampling technique samples a subset of features for each tree. It has been proved to be effective in practice [86] in improving both performance and generalization.

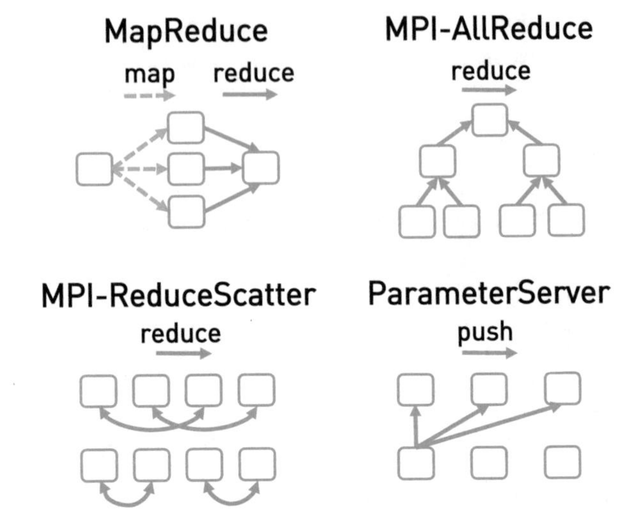

Fig. 3.15 Different parameter sharing architectures of distributed GBDT

3.3.2 *Distributed Gradient Optimization*

When GBDT is trained in a distributed setting, the distributed gradient optimization may use different schemes. From the perspective of parameter sharing structure, it can adopt shared-nothing architecture (e.g., MPI) or shared-memory architecture (e.g., parameter server). While from the perspective of data management, the training data can be partitioned horizontally or vertically.

Parameter Sharing Architectures

Figure 3.15 illustrates different parameter sharing architectures for distributed GBDT, including MapReduce, MPI-AllReduce, MPI-ReduceScatter, and Parameter-Server.

- *MapReduce.* Some distributed solutions, e.g., MLlib [89], choose MapReduce framework for the parameter sharing. Taking Spark as an example of MapReduce, there are two types of nodes: master (driver) and worker (executor). The MapReduce framework partitions the training data horizontally (by rows) over a set of workers. The training process is presented as follows:

  1. In each iteration, the master node pulls off active tree nodes from a waiting queue.
  2. The algorithm splits this set of tree nodes. In order to choose the best split for a given tree node, data statistics (gradients) are collected from distributed workers.
  3. For each tree node, the statistics are collected to a particular worker node via a `reduceByKey` operator.
  4. The designated worker chooses the best split result (feature–value pair) or chooses to stop splitting if the stopping criteria are met.

5. The master collects all decisions about splitting nodes and updates the model.
6. The updated model is passed to the workers on the next iteration.
7. This process iterates until the queue of the active tree node is empty.

In this MapReduce-based framework, statistical aggregation is the bottleneck. In general, this framework is bound either by the cost of statistics computation on workers or by communicating the statistics.

- *MPI-AllReduce.* The AllReduce MPI operator is used in solutions such as XGBoost [86]. The first two steps are the same as the MapReduce processing. Each worker calculates the data statistics (gradients). But different from the MapReduce framework, there is no master node in the AllReduce framework. To perform the aggregation of gradients, MPI-AllReduce uses `AllReduce` operator. All the workers are organized as a binomial tree consisting of leaf workers, internal workers, and a root worker. The statistics of training data are communicated following the bottom-up path of this tree structure. Specifically, the aggregation starts at the leaf worker. In the first communication step, the statistics of two leaf workers are merged on their parent worker. This process repeats in the same manner until reaching the root worker. Once the root worker gets the merged statistics, it calculates the split result and updates the model. The updated model is sent to all the workers via an up-bottom strategy that starts at the root worker and ends at the leaf workers. Then, the next split iteration begins until the whole tree construction is finished.
- *MPI-ReduceScatter.* Another MPI operator, called ReduceScatter, is also applied in some solutions [88]. In distributed training of GBDT, MPI-ReduceScatter operates almost the same as MPI-AllReduce, except for the aggregation stage. Instead of merging statistics via a binomial tree, MPI-ReduceScatter uses the `ReduceScatter` operator. ReduceScatter is a variant of `AllReduce` in which the result, instead of being merged at the root node, is scattered among all workers and each worker is responsible for merging a part of the result. As shown in Fig. 3.15, ReduceScatter uses a recursive halving strategy in which communication cost is halved at each step. Specifically, ReduceScatter sorts all workers as a list. In each communication step, ReduceScatter equally divides the list into two sublists and exchanges necessary statistics within each sublist. The next communication step further divides two sublists. Since the size of a sublist is halved in each step, the communication cost is also halved. This process iterates until the neighboring workers exchange their data.
- *ParameterServer.* When the model size of GBDT becomes large, the parameter server architecture is beneficial because it avoids the single-node bottleneck in MapReduce and MPI-based solutions. Approaches such as TencentBoost [82] and DimBoost [90] choose the parameter server strategy, which partitions the model parameters of GBDT (including tree structures and gradient histogram) over several servers. Each worker has a partition of the training dataset, computes necessary statistics (histograms), and pushes them to the parameter servers. After merging the statistics, the optimal splitting of tree nodes can be found on the parameter servers and sent to the workers.

Given different choices of parameter sharing structures, the question is *how to choose an appropriate structure for a given workload?* Jiang et al. [90] conduct an anatomy of these parameter sharing architectures statistically using a cost model [91]. Assume that there exist w workers and that the size of a local gradient histogram is h bytes. The time needed for a worker to send or receive a package is $\alpha + n\beta$, where α is the latency for each package, β is the transfer time per byte, and n is the number of bytes transferred via the network. γ is the computation cost per byte for merging two histograms. Note that the computation time is often less than the transmission time.

As shown in Fig. 3.15, *MapReduce* merges local parameters in the reduce phase. Since the reduction operation is performed by a single node, the reducing node needs to receive hw bytes. In summary, MLlib takes one communication step and in total $(h\beta w + \alpha + h\gamma)$ time. In *MPI-AllReduce*, all the workers are organized as a binomial tree, and aggregation starts from the leaves and ends at the root—$\log w$ steps in total. Therefore, MPI-AllReduce needs $\log w$ communication steps and $(h\beta + \alpha + h\gamma)\log w$ time. As shown in Fig. 3.15, `MPI-ReduceScatter` scatters the statistic merging among all workers instead of merging at the root node and designs a recursive halving strategy in which the communication cost is halved at each step. Specifically, in the first step, each worker exchanges $\frac{h}{2}$ data with a worker that is $\frac{w}{2}$-distance away; in the second step, each worker exchanges $\frac{h}{4}$ data with a worker that is $\frac{w}{4}$-distance away. This process iterates until the neighboring workers exchange their data. To sum up, MPI-ReduceScatter needs $\log w$ communication steps and $(\frac{w-1}{w}h\beta + (\alpha + h\gamma)\log w)$ time. However, the above result only applies to a case that w is a power of two. If w is not a power of two, the time taken by MPI-ReduceScatter is doubled. For the parameter server architecture, we assume it colocates workers and servers, meaning that there exist one worker and one server on each physical node. Each worker needs to send $(w-1)$ packages in a batch, each of which is $\frac{h}{w}$ bytes. ParameterServer only needs one communication step and takes $(\frac{w-1}{w}h\beta + (w-1)\alpha + h\gamma)$ time.

The time costs of these parameter sharing architectures are summarized in Table 3.3. In the presence of a large histogram, the item with h dominates the overall time cost. It is obvious that Parameter Server and MPI-ReduceScatter outperform the other two with a large w. Therefore, they are more suitable for a large message and a large cluster. As a special case, if w is not a power of two, MPI-ReduceScatter

Table 3.3 Communication costs of different communication architectures for distributed GBDT. w: # workers, h: histogram size, α: latency per package, β: transfer time per byte, γ: merging time per byte

Communication architecture	# comm steps	Communication time
MapReduce	1	$h\beta w + \alpha + h\gamma$
MPI-AllReduce	$\log w$	$(h\beta + \alpha + h\gamma)\log w$
MPI-ReduceScatter	$\log w$	$\frac{w-1}{w}h\beta + (\alpha + h\gamma)\log w$
ParameterServer	1	$\frac{w-1}{w}h\beta + (w-1)\alpha + h\gamma$

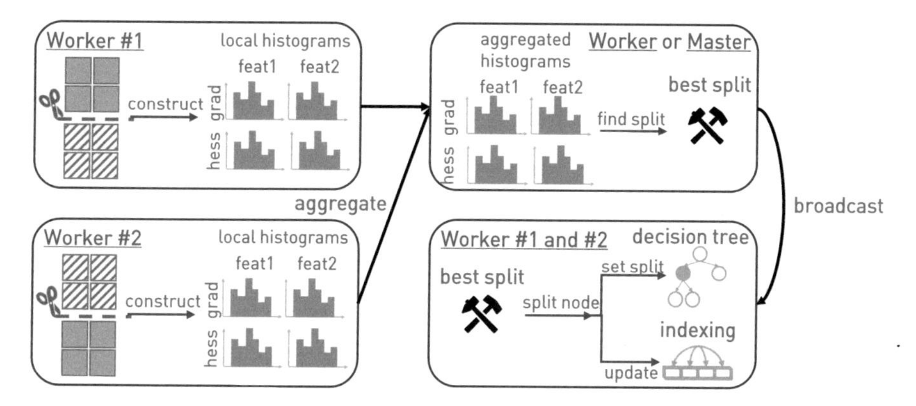

Fig. 3.16 Horizontal partitioning distributed GBDT

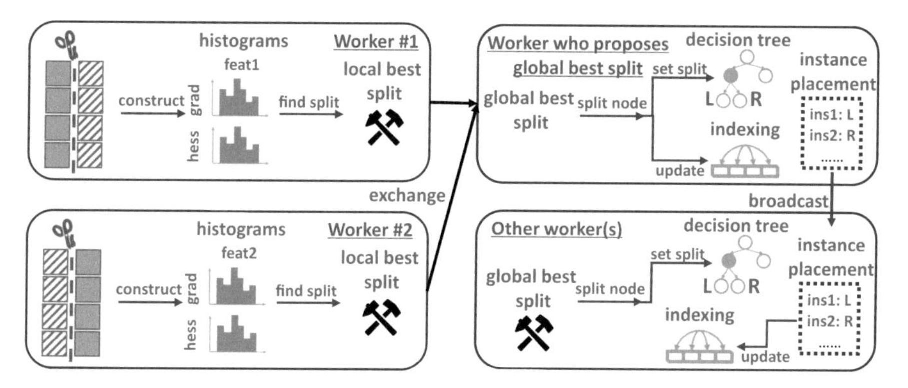

Fig. 3.17 Vertical partitioning distributed GBDT

consumes about twice the time of ParameterServer. In contrast, `MPI-AllReduce` is good for small messages.

Data Management

The parallelism of computation is another fundamental technique in designing distributed machine learning models. Data parallelism is widely adopted in training distributed GBDT in the context of big data. The training dataset is partitioned over several workers, either horizontally or vertically.

Figures 3.16 and 3.17 illustrate the horizontal partitioning and vertical partitioning in training GBDT, respectively.

- *Horizontal partitioning*. The training dataset is horizontally partitioned so that each worker has a subset of instances (rows). To split one tree node, each worker builds local gradient histograms with its local data partition. The local gradient histograms are aggregated on a worker (in MPI or MapReduce) or a master (in parameter server). The best split is calculated with the aggregated gradient histograms. Afterward, the best split is broadcast to all the workers, with which each worker splits the tree node and assigns the instances to the children nodes (either left child or right child). Generally, an index structure is maintained to record the placement of the instances. Horizontal partitioning is adopted in systems such as XGBoost [86] and MLlib [89].

- *Vertical partitioning.* Different from horizontal partitioning, vertical partitioning chooses to partition the training dataset vertically (by columns). With vertical partitioning, each worker has several complete columns (features) of the training data matrix. Since the values of one specific feature can only be found on one worker, there is no need to aggregate the gradient histograms through the network. Each worker obtains the local best split from its feature subset. Then, all workers exchange the local best splits and choose the global best split. Nevertheless, since the feature values of an instance are partitioned, its placement after node splitting, i.e., left or right child node, is only known by the worker that proposes the global best split. As a result, the placement of instances must be broadcast to all workers.

Comparing horizontal partitioning and vertical partitioning, horizontal partitioning needs to communicate the gradient histograms, while vertical partitioning does not have to. Vertical partitioning needs to communicate the split results. Fu et al. [92] theoretically study the trade-off space between horizontal partitioning and vertical partitioning.

The core operation of GBDT is the construction and manipulation of gradient histograms. The size of the histograms is determined by three factors. The first factor is the feature dimension, denoted by D. Since two histograms are built for each feature (one first-order gradient histogram and one second-order gradient histogram), the total size is proportional to $2 \times D$. The second factor is the number of candidate splits, denoted by q. The number of bins in one histogram equals the number of candidate splits, which makes the histogram size proportional to q. The third factor is the number of classes, denoted by C. In multiclassification tasks, the gradient is a vector of partial derivatives on all classes. The histogram size is therefore proportional to C. In summary, the histogram size on one tree node, denoted by $Size_{hist}$, is $2 \times D \times q \times C \times 8$ bytes, where 8 bytes is the size of a double precision floating-point number. The memory cost and communication cost are given below:

- *Memory cost.* The memory cost for both partitioning strategies to store the dataset is similar. Nonetheless, the memory cost to store the gradient histograms is quite different. The maximum number of histograms to be held in memory equals the number of tree nodes in the last but one layer, which is 2^{L-2}, where L denotes the tree depth. With horizontal partitioning, each worker needs to construct the histograms of all features; thus, the memory cost of histograms is $Size_{hist} \times 2^{L-2}$. Nevertheless, with vertical partitioning, each worker constructs the histograms of a portion of features. As a result, the expected memory cost is $Size_{hist} \times \frac{2^{L-2}}{w}$, which is significantly smaller than the horizontal partitioning counterpart.

- *Communication cost.* The communication cost in the horizontal partitioning scheme is mainly the aggregation of histograms. Despite the existence of different aggregation methods [91], such as MapReduce, AllReduce, and ReduceScatter, the minimal transferred data of each worker is the size of the local histograms. Thus, the total communication cost among the cluster for building one tree is at least $Size_{hist} \times w \times (2^{L-1} - 1)$. As the tree goes deeper, i.e., as L increases, the communication cost grows quadratically. Unlike the horizontal partitioning scheme, the vertical partitioning scheme does not need to aggregate the histograms since each worker holds all the values of a specific feature. After splitting a tree node, the placement of instances must be broadcast to all workers. Since the communication cost is only affected by the number of instances, the overhead in one tree layer remains the same as the tree goes deeper. Fu et al. [92] propose to encode the placement into a bitmap so that the communication overhead can be reduced sharply. To conclude, the communication cost for an L-layer tree is $\lceil \frac{N}{8} \rceil \times w \times L$ bytes, where $\lceil \frac{N}{8} \rceil$ bytes is the size of one bitmap.

The choice of partitioning scheme highly depends on the size of gradient histogram $Size_{hist}$. Undoubtedly, horizontal partitioning works well for datasets with low dimensionality since the resulting histograms are small. However, in both industry and academia, the following three cases become increasingly popular—high-dimensional features, deep trees, and multiclassification. In these cases, the histogram size can be very large. Therefore, vertical partitioning is far more memory- and communication-efficient than horizontal partitioning. Take an industrial dataset Age as an example [82], the dataset contains 48M instances, 0.33 million features, and 9 classes. Assuming each tree has 8 layers and the number of candidate splits is 20, we train a GBDT model over 8 workers. The estimated size of histograms on one tree node can be up to 906 MB. Using the horizontal partitioning approach, the memory consumption would be 56.6 GB and the total communication cost would be 900 GB for merely one tree in the worst case. On the contrary, when the vertical scheme is applied, the expected memory cost of histograms is 7.08 GB per tree and the communication cost is merely 366 MB for one tree.

Although vertical partitioning can be superior for high-dimensional datasets, an issue in the implementation of vertical partitioning is how to partition the input training dataset. Since the original datasets are often stored by rows, it is nontrivial to transform the original dataset to vertical partitions. This can be implemented through a MapReduce framework [92].

References

1. Nelder, John Ashworth and Wedderburn, Robert WM: Generalized linear models. Journal of the Royal Statistical Society: Series A (General). 135(3), 370–384 (1972)
2. Emily Fox: CSE446: Machine Learning, University of Washington. https://courses.cs.washington.edu/courses/cse446/ (2020)

3. Shalev-Shwartz, Shai and Tewari, Ambuj: Stochastic methods for l1-regularized loss minimization. The Journal of Machine Learning Research. 12, 1865–1892 (2011)
4. Meier, Lukas and Van De Geer, Sara and Bühlmann, Peter: The group lasso for logistic regression. Journal of the Royal Statistical Society: Series B (Statistical Methodology). 70(1), 53–71 (2008)
5. Tsuruoka, Yoshimasa and Tsujii, Junichi and Ananiadou, Sophia: Stochastic gradient descent training for l1-regularized log-linear models with cumulative penalty. Proceedings of the Joint Conference of the 47th Annual Meeting of the ACL and the 4th International Joint Conference on Natural Language Processing of the AFNLP. 477–485 (2009)
6. Carpenter, Bob: Lazy sparse stochastic gradient descent for regularized multinomial logistic regression. Alias-i, Inc., Tech. Rep. 1–20 (2008)
7. Blei, David M: Linear regression, Logistic regression, and Generalized Linear Models. (2014)
8. Jurafsky, Dan: Speech & language processing. Pearson Education India. (2000)
9. Böhning, Dankmar: Multinomial logistic regression algorithm. Annals of the Institute of Statistical Mathematics. 44(1), 197–200 (1992)
10. Noble, William S: What is a support vector machine?. Nature Biotechnology. 24(12), 1565–1567 (2006)
11. Suykens, Johan AK and Vandewalle, Joos: Least squares support vector machine classifiers. Neural processing letters. 9(3), 293–300 (1999)
12. SVM margin, https://math.stackexchange.com/questions/1305925/why-is-the-svm-margin-equal-to-frac2-mathbfw (2020)
13. Joachims, Thorsten: Svmlight: Support vector machine. SVM-Light Support Vector Machine http://svmlight.joachims.org/, University of Dortmund. 19(4) (1999)
14. Recht, Benjamin and Re, Christopher and Wright, Stephen and Niu, Feng: Hogwild: A lock-free approach to parallelizing stochastic gradient descent. Advances in Neural Information Processing Systems. 693–701 (2011)
15. Zinkevich, Martin and Langford, John and Smola, Alex J: Slow learners are fast. Advances in Neural Information Processing Systems. 2331–2339 (2009)
16. Zhang, Ce and Ré, Christopher: Dimmwitted: A study of main-memory statistical analytics. Proceedings of the VLDB Endowment. 7(12), 1283–1294 (2014)
17. Wright, Stephen J: Coordinate descent algorithms. Mathematical Programming. 151(1), 3–34 (2015)
18. Low, Yucheng and Gonzalez, Joseph and Kyrola, Aapo and Bickson, Danny and Guestrin, Carlos and Hellerstein, Joseph M: Distributed GraphLab: A Framework for Machine Learning and Data Mining in the Cloud. Proceedings of the VLDB Endowment. 5(8) (2012)
19. Carter, Chris K and Kohn, Robert: On Gibbs sampling for state space models. Biometrika. 81(3), 541–553 (1994)
20. Zaharia, Matei and Xin, Reynold S and Wendell, Patrick and Das, Tathagata and Armbrust, Michael and Dave, Ankur and Meng, Xiangrui and Rosen, Josh and Venkataraman, Shivaram and Franklin, Michael J and others: Apache spark: a unified engine for big data processing. Communications of the ACM. 59(11), 56–65 (2016)
21. Chu, Cheng-Tao and Kim, Sang and Lin, Yi-An and Yu, YuanYuan and Bradski, Gary and Olukotun, Kunle and Ng, Andrew: Map-reduce for machine learning on multicore. Advances in neural information processing systems. 19, 281–288 (2006)
22. Zinkevich, Martin and Weimer, Markus and Li, Lihong and Smola, Alex J: Parallelized stochastic gradient descent. Advances in Neural Information Processing Systems. 2595–2603 (2010)
23. Polyak, Boris T and Juditsky, Anatoli B: Acceleration of stochastic approximation by averaging. SIAM Journal on Control and Optimization. 30(4), 838–855 (1992)
24. Mcdonald, Ryan and Mohri, Mehryar and Silberman, Nathan and Walker, Dan and Mann, Gideon: Efficient large-scale distributed training of conditional maximum entropy models. Advances in Neural Information Processing Systems. 22, 1231–1239 (2009)

25. Li, Mu and Andersen, David G and Park, Jun Woo and Smola, Alexander J and Ahmed, Amr and Josifovski, Vanja and Long, James and Shekita, Eugene J and Su, Bor-Yiing: Scaling distributed machine learning with the parameter server. 11th USENIX Symposium on Operating Systems Design and Implementation (OSDI 14). 583–598 (2014)
26. Xing, Eric P and Ho, Qirong and Dai, Wei and Kim, Jin Kyu and Wei, Jinliang and Lee, Seunghak and Zheng, Xun and Xie, Pengtao and Kumar, Abhimanu and Yu, Yaoliang: Petuum: A new platform for distributed machine learning on big data. IEEE Transactions on Big Data. 1(2), 49–67 (2015)
27. Jiang, Jiawei and Cui, Bin and Zhang, Ce and Yu, Lele: Heterogeneity-aware distributed parameter servers. Proceedings of the 2017 ACM International Conference on Management of Data. 463–478 (2017)
28. Dean, Jeffrey and Corrado, Greg and Monga, Rajat and Chen, Kai and Devin, Matthieu and Mao, Mark and Ranzato, Marc'aurelio and Senior, Andrew and Tucker, Paul and Yang, Ke and others: Large scale distributed deep networks. Advances in Neural Information Processing Systems. 1223–1231 (2012)
29. Ho, Qirong and Cipar, James and Cui, Henggang and Lee, Seunghak and Kim, Jin Kyu and Gibbons, Phillip B and Gibson, Garth A and Ganger, Greg and Xing, Eric P: More effective distributed ml via a stale synchronous parallel parameter server. Advances in Neural Information Processing Systems. 1223–1231 (2013)
30. Zhang, Zhipeng and Wu, Wentao and Jiang, Jiawei and Yu, Lele and Cui, Bin and Zhang, Ce: ColumnSGD: A Column-oriented Framework for Distributed Stochastic Gradient Descent. 2020 IEEE 36th International Conference on Data Engineering (ICDE). 1513–1524 (2020)
31. Ordentlich, Erik and Yang, Lee and Feng, Andy and Cnudde, Peter and Grbovic, Mihajlo and Djuric, Nemanja and Radosavljevic, Vladan and Owens, Gavin: Network-efficient distributed word2vec training system for large vocabularies. Proceedings of the 25th ACM International on Conference on Information and Knowledge Management. 1139–1148 (2016)
32. Mikolov, Tomas and Sutskever, Ilya and Chen, Kai and Corrado, Greg S and Dean, Jeff: Distributed representations of words and phrases and their compositionality. Advances in neural information processing systems. 26, 3111–3119 (2013)
33. Kara, Kaan and Eguro, Ken and Zhang, Ce and Alonso, Gustavo: ColumnML: Column-Store Machine Learning with On-The-Fly Data Transformation. Proceedings of the VLDB Endowment. 26(4) (2019)
34. Cui, Henggang and Cipar, James and Ho, Qirong and Kim, Jin Kyu and Lee, Seunghak and Kumar, Abhimanu and Wei, Jinliang and Dai, Wei and Ganger, Gregory R and Gibbons, Phillip B and others: Exploiting bounded staleness to speed up big data analytics. 2014 USENIX Annual Technical Conference (USENIX ATC 14). 37–48 (2014)
35. Dai, Wei and Kumar, Abhimanu and Wei, Jinliang and Ho, Qirong and Gibson, Garth and Xing, Eric P: High-performance distributed ML at scale through parameter server consistency models. arXiv preprint arXiv:1410.8043. (2014)
36. Zhang, Zhipeng and Jiang, Jiawei and Wu, Wentao and Zhang, Ce and Yu, Lele and Cui, Bin: Mllib*: Fast training of GLMs using spark Mllib. 2019 IEEE 35th International Conference on Data Engineering (ICDE). 1778–1789 (2019)
37. Maas, Andrew L and Hannun, Awni Y and Ng, Andrew Y: Rectifier nonlinearities improve neural network acoustic models. ICML Workshop on Deep Learning for Audio, Speech and Language Processing. 30(1), 3 (2013)
38. Rumelhart, David E and Hinton, Geoffrey E and Williams, Ronald J: Learning internal representations by error propagation. California Univ San Diego La Jolla Inst for Cognitive Science.
39. Fan, Jianqing and Ma, Cong and Zhong, Yiqiao: A selective overview of deep learning. arXiv preprint arXiv:1904.05526. (2019)
40. Du, Xuedan and Cai, Yinghao and Wang, Shuo and Zhang, Leijie: Overview of deep learning. 2016 31st Youth Academic Annual Conference of Chinese Association of Automation (YAC). 159–1643 (2016)

41. Schmidhuber, Jürgen: Deep learning in neural networks: An overview. Neural networks. 61, 85–117 (2015)
42. Ng, Andrew and others: Sparse autoencoder. CS294A Lecture notes. 72, 1–19 (2011)
43. Vincent, Pascal and Larochelle, Hugo and Bengio, Yoshua and Manzagol, Pierre-Antoine: Extracting and composing robust features with denoising autoencoders. Proceedings of the 25th international conference on Machine learning. 1096–1103 (2008)
44. Hinton, Geoffrey E and Osindero, Simon and Teh, Yee-Whye: A fast learning algorithm for deep belief nets. Neural computation. 18(7), 1527–1554 (2006)
45. Liu, Ping and Han, Shizhong and Meng, Zibo and Tong, Yan: Facial expression recognition via a boosted deep belief network. Proceedings of the IEEE Conference on Computer Vision and Pattern Recognition. 1805–1812 (2014)
46. Kim, Soowoong and Park, Bogun and Song, Bong Seop and Yang, Seungjoon: Deep belief network-based statistical feature learning for fingerprint liveness detection. Pattern Recognition Letters. 77, 58–65 (2016)
47. Smolensky, P: Parallel distributed processing: Explorations in the microstructure of cognition, vol. 1. Chapter Information Processing in Dynamical Systems: Foundations of Harmony Theory. MIT Press. 194–281 (1986)
48. LeCun, Yann and Bottou, Léon and Bengio, Yoshua and Haffner, Patrick: Gradient-based learning applied to document recognition. Proceedings of the IEEE. 86 (11), 2278–2324 (1998)
49. Krizhevsky, Alex and Sutskever, Ilya and Hinton, Geoffrey E: Imagenet classification with deep convolutional neural networks. Communications of the ACM. 60 (6), 84–90 (2017)
50. He, Kaiming and Zhang, Xiangyu and Ren, Shaoqing and Sun, Jian: Deep residual learning for image recognition. Proceedings of the IEEE Conference on Computer Vision and Pattern Recognition. 770–778 (2016)
51. Simonyan, Karen and Zisserman, Andrew: Very deep convolutional networks for large-scale image recognition. arXiv preprint arXiv:1409.1556. (2014)
52. Szegedy, Christian and Liu, Wei and Jia, Yangqing and Sermanet, Pierre and Reed, Scott and Anguelov, Dragomir and Erhan, Dumitru and Vanhoucke, Vincent and Rabinovich, Andrew: Going deeper with convolutions. Proceedings of the IEEE Conference on Computer Vision and Pattern Recognition. 1–9 (2014)
53. Cho, Kyunghyun and Van Merriënboer, Bart and Gulcehre, Caglar and Bahdanau, Dzmitry and Bougares, Fethi and Schwenk, Holger and Bengio, Yoshua: Learning phrase representations using RNN encoder-decoder for statistical machine translation. arXiv preprint arXiv:1406.1078. (2014)
54. Hochreiter, Sepp and Schmidhuber, Jürgen: Long short-term memory. Neural computation. 9(8), 1735–1780 (1997)
55. Sak, Hasim and Senior, Andrew W and Beaufays, Françoise: Long short-term memory recurrent neural network architectures for large-scale acoustic modeling. (2014)
56. Goodfellow, Ian and Pouget-Abadie, Jean and Mirza, Mehdi and Xu, Bing and Warde-Farley, David and Ozair, Sherjil and Courville, Aaron and Bengio, Yoshua: Generative adversarial nets. Advances in Neural Information Processing Systems. 2672–2680 (2014)
57. Mnih, Volodymyr and Kavukcuoglu, Koray and Silver, David and Graves, Alex and Antonoglou, Ioannis and Wierstra, Daan and Riedmiller, Martin: Playing Atari with deep reinforcement learning. arXiv preprint arXiv:1312.5602. (2013)
58. Lian, Xiangru and Zhang, Ce and Zhang, Huan and Hsieh, Cho-Jui and Zhang, Wei and Liu, Ji: Can decentralized algorithms outperform centralized algorithms? a case study for decentralized parallel stochastic gradient descent. Advances in Neural Information Processing Systems. 5330–5340 (2017)
59. Lan, Guanghui and Lee, Soomin and Zhou, Yi: Communication-efficient algorithms for decentralized and stochastic optimization. Mathematical Programming. 180(1), 237–284 (2020)

60. Sirb, Benjamin and Ye, Xiaojing: Consensus optimization with delayed and stochastic gradients on decentralized networks. 2016 IEEE International Conference on Big Data (Big Data). 76–85 (2016)
61. Ram, S Sundhar and Nedić, A and Veeravalli, Venugopal V: Asynchronous gossip algorithms for stochastic optimization. Proceedings of the 48th IEEE Conference on Decision and Control (CDC) held jointly with the 2009 28th Chinese Control Conference. 3581–3586 (2009)
62. Srivastava, Kunal and Nedic, Angelia: Distributed asynchronous constrained stochastic optimization. IEEE Journal of Selected Topics in Signal Processing. 5(4), 772–790 (2011)
63. Ram, S Sundhar and Nedić, Angelia and Veeravalli, Venugopal V: Distributed stochastic subgradient projection algorithms for convex optimization. Journal of optimization theory and applications. 147(3), 516–545 (2010)
64. Lian, Xiangru and Zhang, Wei and Zhang, Ce and Liu, Ji: Asynchronous decentralized parallel stochastic gradient descent. International Conference on Machine Learning. 3043–3052 (2018)
65. Alistarh, Dan and Grubic, Demjan and Li, Jerry and Tomioka, Ryota and Vojnovic, Milan: QSGD: Communication-efficient SGD via gradient quantization and encoding. Advances in Neural Information Processing Systems. 1709–1720 (2017)
66. Elias, Peter: Universal codeword sets and representations of the integers. IEEE Transactions on Information Theory. 21(2), 194–203 (1975)
67. Seide, Frank and Fu, Hao and Droppo, Jasha and Li, Gang and Yu, Dong: 1-bit stochastic gradient descent and its application to data-parallel distributed training of speech DNNs. Fifteenth Annual Conference of the International Speech Communication Association. (2014)
68. Wu, Jiaxiang and Huang, Weidong and Huang, Junzhou and Zhang, Tong: Error compensated quantized sgd and its applications to large-scale distributed optimization. 80, 5325 (2018)
69. Jiang, Jiawei and Fu, Fangcheng and Yang, Tong and Cui, Bin: Sketchml: Accelerating distributed machine learning with data sketches. Proceedings of the 2018 International Conference on Management of Data. 1269–1284 (2018)
70. Greenwald, Michael and Khanna, Sanjeev: Space-efficient online computation of quantile summaries. ACM SIGMOD Record. 30(2), 58–66 (2001)
71. Zhang, Qi and Wang, Wei: A fast algorithm for approximate quantiles in high speed data streams. 19th International Conference on Scientific and Statistical Database Management (SSDBM 2007). 29–29 (2007)
72. Data Sketches, https://datasketches.github.io/
73. Cormode, Graham and Muthukrishnan, Shan: An improved data stream summary: the count-min sketch and its applications. Journal of Algorithms. 55(1), 258–75 (2005)
74. Aji, Alham Fikri and Heafield, Kenneth: Sparse communication for distributed gradient descent. arXiv preprint arXiv:1704.05021. (2017)
75. Garg, Rahul and Khandekar, Rohit: Gradient descent with sparsification: an iterative algorithm for sparse recovery with restricted isometry property. Proceedings of the 26th Annual International Conference on Machine Learning. 337–344 (2009)
76. Wang, Hongyi and Sievert, Scott and Liu, Shengchao and Charles, Zachary and Papailiopoulos, Dimitris and Wright, Stephen: Atomo: Communication-efficient learning via atomic sparsification. Advances in Neural Information Processing Systems. 9850–9861 (2018)
77. Stich, Sebastian U and Cordonnier, Jean-Baptiste and Jaggi, Martin: Sparsified SGD with memory. Advances in Neural Information Processing Systems. 4447–4458 (2018)
78. Lin, Yujun and Han, Song and Mao, Huizi and Wang, Yu and Dally, William J: Deep gradient compression: Reducing the communication bandwidth for distributed training. arXiv preprint arXiv:1712.01887. (2017)
79. Alistarh, Dan and Hoefler, Torsten and Johansson, Mikael and Konstantinov, Nikola and Khirirat, Sarit and Renggli, Cédric: The convergence of sparsified gradient methods. Advances in Neural Information Processing Systems. 5973–5983 (2018)

80. Wangni, Jianqiao and Wang, Jialei and Liu, Ji and Zhang, Tong: Gradient sparsification for communication-efficient distributed optimization. Advances in Neural Information Processing Systems. 31, 1299–1309 (2018)
81. Wang, Wei and Chen, Gang and Dinh, Anh Tien Tuan and Gao, Jinyang and Ooi, Beng Chin and Tan, Kian-Lee and Wang, Sheng: SINGA: Putting deep learning in the hands of multimedia users. Proceedings of the 23rd ACM International Conference on Multimedia. 25–34 (2015)
82. Jiang, Jie and Jiang, Jiawei and Cui, Bin and Zhang, Ce: Tencentboost: A gradient boosting tree system with parameter server. 2017 IEEE 33rd International Conference on Data Engineering (ICDE). 281–284 (2017)
83. Friedman, Jerome H: Greedy function approximation: a gradient boosting machine. Annals of statistics. 1189–1232 (2001)
84. Breiman, Leo: Random forests. Machine learning. 45(1), 5–32 (2001)
85. Bottou, Léon: Large-scale machine learning with stochastic gradient descent. Proceedings of COMPSTAT'2010. 177–186 (2010)
86. Chen, Tianqi and Guestrin, Carlos: Xgboost: A scalable tree boosting system. Proceedings of the 22nd ACM SIGKDD International Conference on Knowledge Discovery and Data Mining. 785–794 (2016)
87. Friedman, Jerome and others: Additive logistic regression: a statistical view of boosting. Annals of statistics. 28(3), 337–407 (2000)
88. Meng, Qi and Ke, Guolin and Wang, Taifeng and Chen, Wei and Ye, Qiwei and Ma, Zhi-Ming and Liu, Tieyan: A communication-efficient parallel algorithm for decision tree. Advances in Neural Information Processing Systems. 1279–1287 (2016)
89. Meng, Xiangrui and Bradley, Joseph and Yavuz, Burak and Sparks, Evan and Venkataraman, Shivaram and Liu, Davies and Freeman, Jeremy and Tsai, DB and Amde, Manish and Owen, Sean and others: Mllib: Machine learning in apache spark. The Journal of Machine Learning Research. 17(1), 1235–1241 (2016)
90. Jiang, Jiawei and Cui, Bin and Zhang, Ce and Fu, Fangcheng: Dimboost: Boosting gradient boosting decision tree to higher dimensions. Proceedings of the 2018 International Conference on Management of Data. 1363–1376 (2018)
91. Thakur, Rajeev and Rabenseifner, Rolf and Gropp, William: Optimization of collective communication operations in MPICH. The International Journal of High Performance Computing Applications. 19(1), 49–66 (2005)
92. Fu, Fangeheng and Jiang, Jiawei and Shao, Yingxia and Cui, Bin: An experimental evaluation of large scale GBDT systems. Proceedings of the VLDB Endowment. 12(11), 1357–1370 (2019)

Chapter 4
Distributed Machine Learning Systems

Abstract In recent decades, many distributed machine learning systems have been designed in both academia and industry. They are proposed for different purposes and adopt different architectures. Some are general systems that can run different kinds of ML models through delicate systematic abstractions, while others focus on a specific class of ML models through highly specific optimizations. Besides, machine learning systems are constructed over different underlying infrastructures, e.g., new hardware (GPU/FPGA/RDMA), cloud environment, and databases. In this chapter, we will describe a broad range of machine learning systems in terms of motivations, architectures, functionalities, pros, and cons.

4.1 General Machine Learning Systems

The diversity of machine learning models motivates the demand for general machine learning systems that can efficiently handle different kinds of machine learning models under a unified framework. Since different machine learning models may adopt different training schemes, building such general machine learning models requires high-level abstractions of parallelism and programming. We summarize the existing machine learning system into three categories according to their underlying architectures—MapReduce systems, parameter server systems, and specialized systems.

4.1.1 MapReduce Systems

MapReduce has been one of the most popular parallel computing frameworks in the last decades due to its simple, yet effective, abstraction and efficient performance [1]. Typical MapReduce frameworks include Hadoop [2] and Spark [3, 4]. The whole MapReduce process takes key-value pairs as the input and outputs other key-value pairs. There are two programming abstractions—*map* and *reduce*.

J. Jiang et al., *Distributed Machine Learning and Gradient Optimization*, Big Data Management, https://doi.org/10.1007/978-981-16-3420-8_4

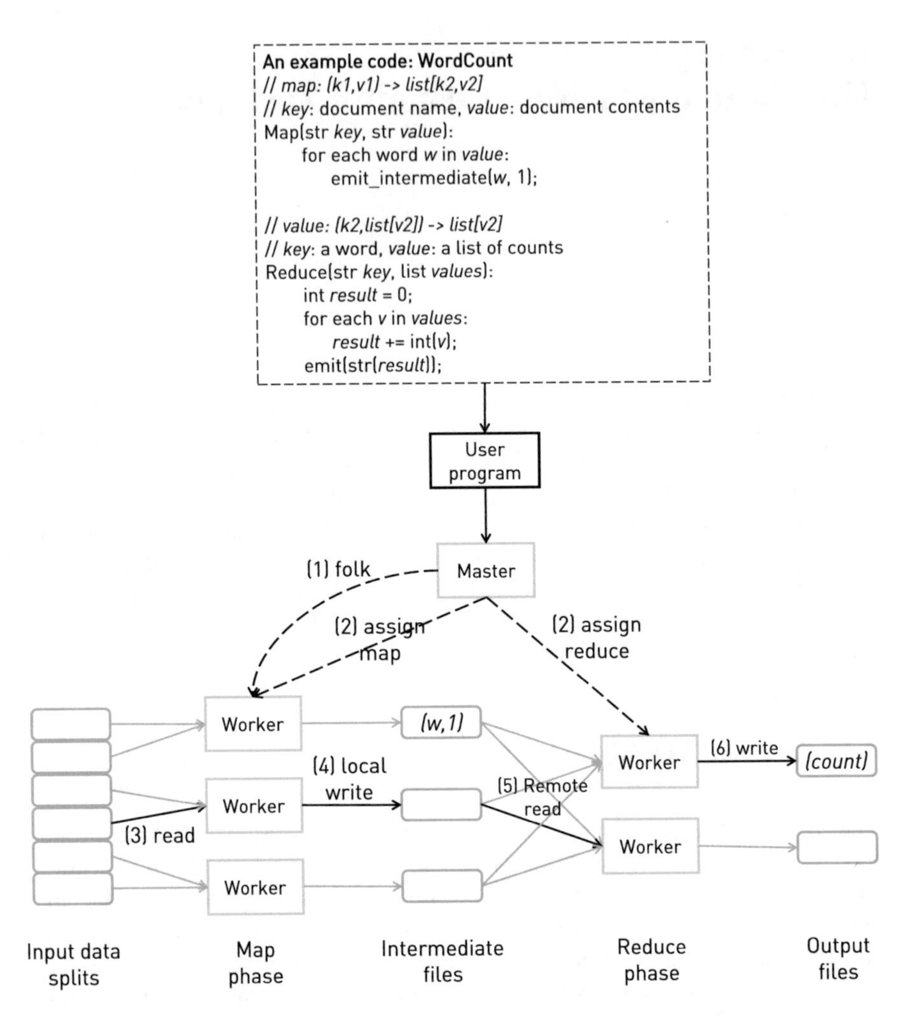

Fig. 4.1 Execution flow of MapReduce

map: (k1,v1) → *list[k2,v2]* function takes a key-value pair as input and outputs a list of intermediate key-value pairs. Then, the pairs with the same intermediate key are grouped together as the input for the downstream *reduce* function. *reduce: (k2,list[v2])* → *list[v2]* function takes an input of an intermediate key and a list of corresponding values. It uses some merging function to merge these values into another list of values. Generally, the output of *reduce* has zero or one value. Figure 4.1 illustrates an example of WordCount, including the programming and execution flow.

1. The input files, typically stored in a distributed file system such as HDFS, are partitioned into several splits. The split size can be configured by the user, typically 16 MB to 64 MB. Each worker handles one or several data splits.

2. The master node is responsible for assigning *map* and *reduce* tasks to idle workers.
3. During the *map* phase, each assigned worker reads its corresponding data split. The worker reads each input key-value pair from the data split and then uses the user-defined function to process each pair. The produced intermediate key-value pairs are buffered in memory.
4. The intermediate key-value pairs are written to the local disk periodically according to the partitioning strategy.
5. After the *map* phase is finished, the reducing workers assigned by the master use remote procedure calls to read the intermediate key-value pairs from the *map* workers. Once all the responsible intermediate data are obtained, each *reduce* worker groups the values of the same intermediate key. This generally requires sorting the intermediate pairs.
6. Each reducing worker iterates the grouped intermediate data. For each intermediate key, it leverages the user-defined to process the corresponding set of values and generate the outputs.

The *map* and *reduce* abstractions of MapReduce make it easy to design parallel algorithms. In the area of distributed machine learning, several systems choose MapReduce as their fundamental infrastructure, such as Mahout and MLlib.

Mahout

Apache Mahout [5] is an open-source project developed by Apache Software Foundation (ASF) for the goal of creating scalable machine learning algorithms (focused on linear algebra and statistics) that are efficient in execution and easy-to-use for the users. Mahout chooses the Hadoop library as its underlying engine and implements a wide range of machine learning models, e.g., classification, regression, recommendation, and clustering.

The logical architecture and physical architecture of Mahout are shown in Fig. 4.2. The logical architecture of Mahout can be decomposed into four layers:

- *Storage layer.* Mahout reads and writes data from/to a distributed file system, generally HDFS in the Hadoop ecosystem.
- *Computation layer.* The computation layer provides basic computation units, such as utilities from Apache Lucene, math library, collection library, and Hadoop library.
- *Algorithm layer.* A broad spectrum of machine learning models is implemented using Mahout and provided to the users.
- *Application layer.* Practitioners can build their applications using the algorithms in Mahout.

The physical architecture of Mahout depends on MapReduce. There are a master node and several slave nodes. The master node runs a resource manager and a name node to schedule the slaves and manage the locations of data splits.

Figure 4.3 showcases the execution of KMeans in Mahout. Initially, the map phase lets each map node read responsible data split from HDFS. Each map node uses the same initial centroids and assigns each data point to one centroid. Then,

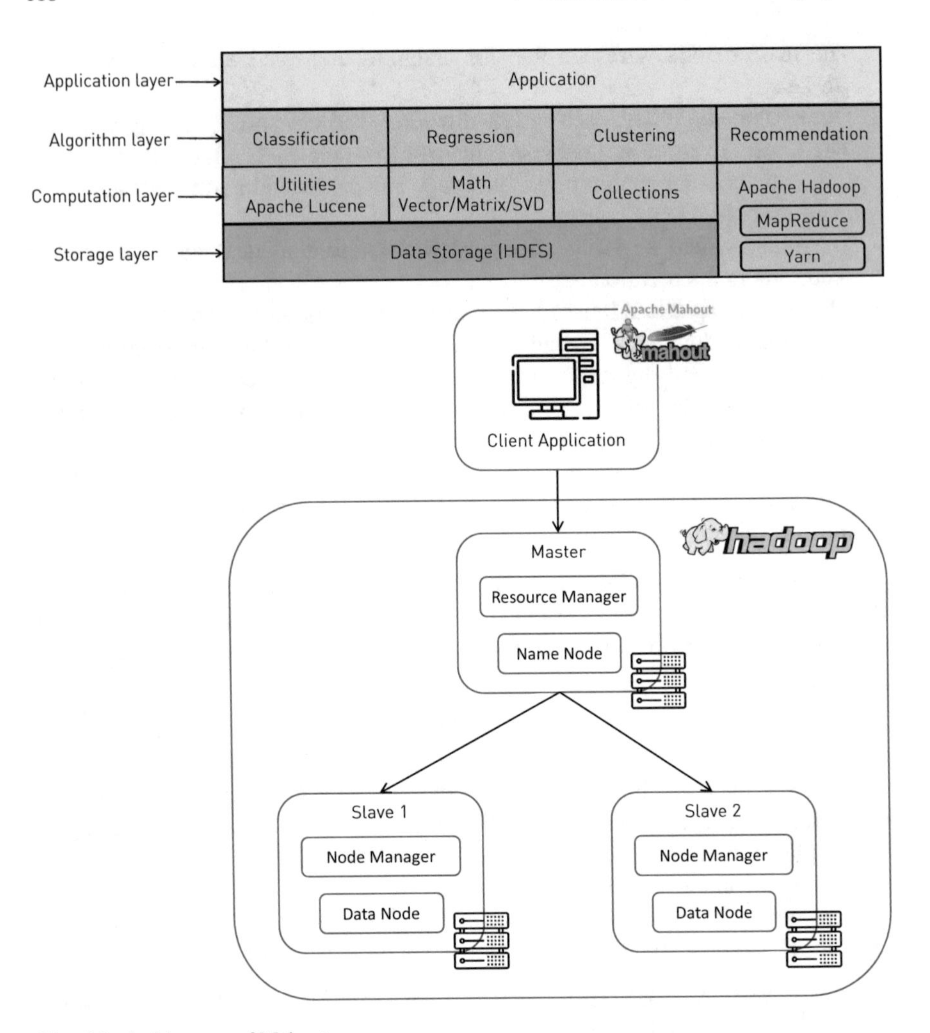

Fig. 4.2 Architecture of Mahout

the intermediate *(centroid_id, data_point)* pairs are stored in the local disk. When the map phase is finished, the reduce phase groups the intermediate pairs by their keys. Since the number of centroids is k, there are k reducing nodes to perform the reduce functionality. Each reduce node computes the new centroid with the data points belonging to the corresponding centroid. This distributed training scheme is able to generalize to other kinds of machine learning models. For example, Logistic Regression can be similarly implemented by computing the gradients in the map phase and updating the model in the reduce phase.

Note that since the release of 0.10.0, Mahout started migrating from Hadoop to Spark. Mahout designs a programming environment called Samsara and a Scala DSL supporting distributed algebraic operations.

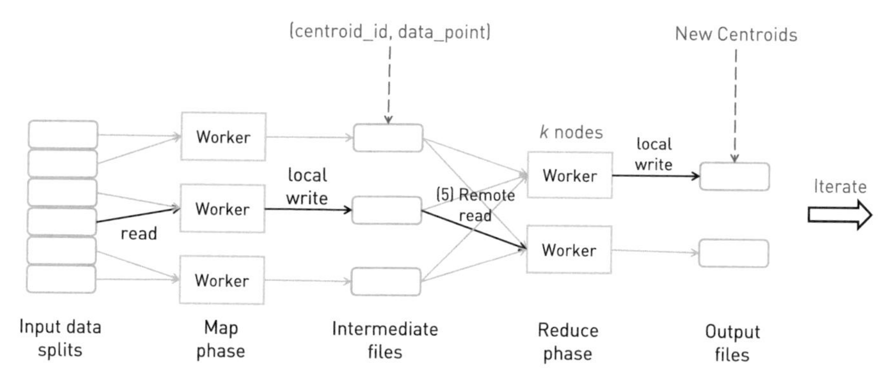

Fig. 4.3 Implementation of KMeans in Mahout

Mllib

MLlib [6] is a scalable machine learning library in Apache Spark [4]. MLlib provides multilanguage programming interfaces for users, such as Scala, Java, Python, and R. Compared with Hadoop-based machine learning systems, MLlib is significantly faster because Spark outperforms Hadoop. Besides, MLlib is closely integrated with many open-source projects, e.g., Yarn, Mesos, and Kubernetes, so that it is easy for the user to develop applications with MLlib.

Before presenting MLlib, we first introduce its underlying computation engine Spark. Although Hadoop has delicately designed primitives for parallel computing, it reveals several limitations facing large volumes of data. First, *map* and *reduce* are low-level programming abstractions and therefore it is often difficult for the users to directly develop complex algorithms. Second, Hadoop needs to frequently write and read data to the local disk. For large-scale applications, Hadoop may encounter a serious performance bottleneck. Third, Hadoop is inefficient for many iterative machine learning models. These models need to repeatedly read the same dataset during the training. Each iteration, as a MapReduce job, must load the dataset from the disk, causing expensive overheads.

Architecture of Spark To address the bottlenecks of Hadoop, Zaharia et al. design a new parallel computing framework called Spark [4]. The architecture of Spark is shown in Fig. 4.4. A Spark application runs with separate resources on a cluster, including a driver and several executors. An object called *SparkContext* runs in the driver which coordinates the whole job. The *SparkContext* communicates with a cluster manager (e.g., Spark's cluster manager, Yarn, or Mesos) and allocates resources for the application. Once the resources are allocated, the executors are launched, and the SparkContext sends tasks to the executors to run according to the application code from the user.

RDD The main abstraction of the data structure in Spark is called resilient distributed dataset (RDD), which is a read-only object of collections. RDD is partitioned into several partitions and stored over several machines. The partitions

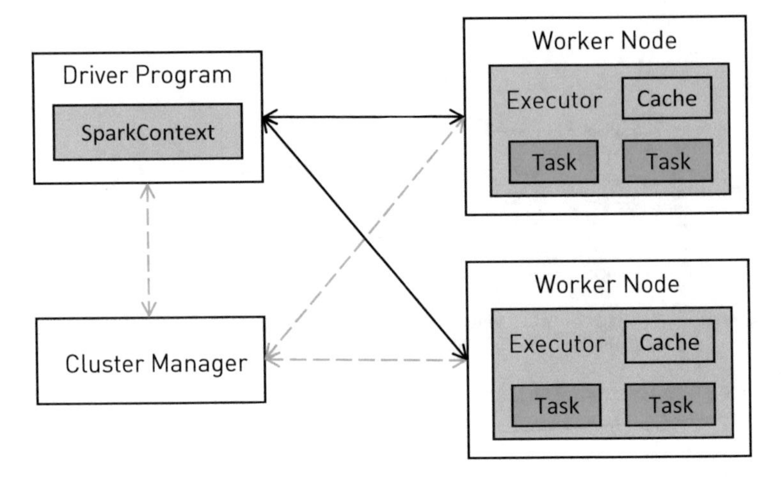

Fig. 4.4 Architecture of Spark

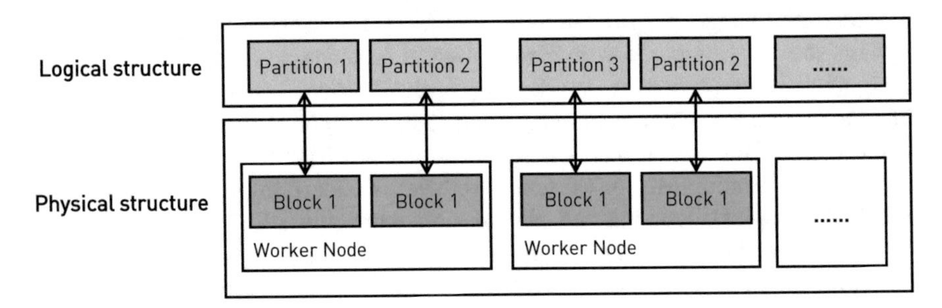

Fig. 4.5 RDD data structure

can be processed in parallel using various parallel operators, as shown in Fig. 4.5. In terms of logical structure, an RDD consists of several partitions, while in terms of physical structure, each worker node has one or several partitions. The parallelism of RDD partitions determines the parallelism of the running job. An RDD can be cached in memory, which can be reused in iterative MapReduce jobs. In this way, Spark avoids frequently reading and writing intermediate data from/to the disk. RDD can be created in two ways—parallelizing an existing collection in the driver program or parsing a dataset from an external storage, such as local disk, HDFS, and HBase. Spark offers two types of operations for RDD—*transformations* and *actions*, as shown in Fig. 4.6.

- *Transformations.* Transformations create a new RDD of type A using an existing RDD of type B. Typical transformation operators include *map*, **flatMap**, *filter*, *union*, and *join*. For example, *flatMap* is a transformation operator that transforms each element of type A to $List[B]$. Transformation operators are lazy, meaning that the upstream transformations over a base RDD are remembered, and these transformation operators are actually executed when an action operation is

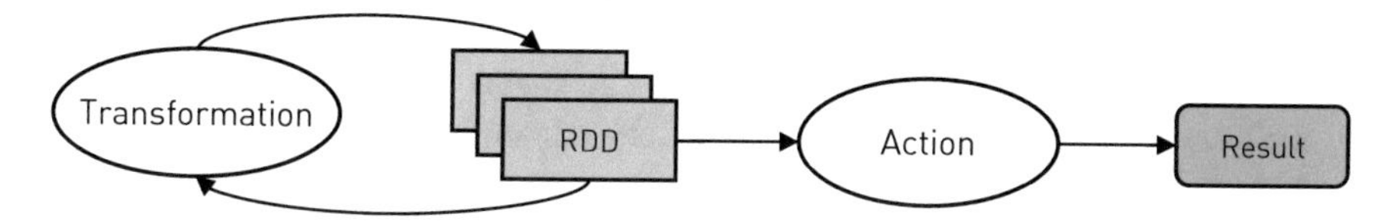

Fig. 4.6 Transformation and action operators of RDD

required. This lazy execution is more efficient by merging some transformations together and save unnecessary communications.

- *Actions.* Actions return a value to driver program by computing on an RDD, such as *count*, *reduce*, and *reduceByKey*. For instance, reduce is an action operator that aggregates all the elements in the RDD using the user-defined function to a single value and returns the final result to the driver program.

Failure Recovery A Spark application may generate different RDDs through transformation operators. These RDDs can be represented as a DAG (Directed Acyclic Graph), in which each vertex denotes an RDD and each edge denotes an operator. When the running Spark application encounters failure (e.g., a partition of RDD is lost), Spark can reconstruct the lost partitions of RDD using lineage information represented by the RDD DAG. An extreme case is reconstructing all the previous RDDs from to original data source. To avoid the cost of failure recovery, Spark provides functionality to periodically store checkpoints on storage such as HDFS.

Persistence and Partitioning Users can choose different storage strategies, such as *MEMORY_ONLY*, *MEMORY_AND_DISK*, and *DISK_ONLY*, to persist RDDs into memory or disk. User can also partition an existing RDD using various partitioning strategies, e.g., hash partitioning, 2D partitioning, and 3D partitioning.

Architecture of MLlib MLlib [6] is a subproject of Spark which focuses on scalable machine learning. Figure 4.7 shows the ecosystem of Spark, in which MLlib, Spark SQL, Spark Streaming, and GraphX are submodules built on top of Spark.

- *Supported Machine Learning Models.* MLlib supports a wide range of machine learning models, including but not limited to classification models (logistic regression, support vector machine, naive Bayes), regression models (linear regression, Lasso), clustering models (KMeans, Gaussian mixtures), recommen-

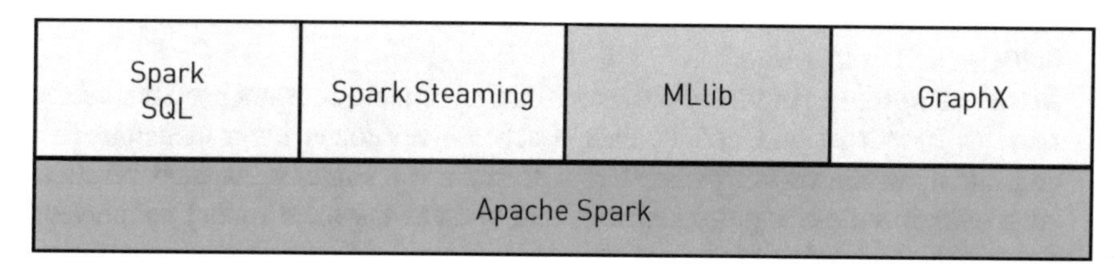

Fig. 4.7 Apache Spark ecosystem

dation models, tree models (decision tree, random forest, gradient boosting decision tree), and dimension reduction (PCA, SVD).

- *Utilities.* In addition to the training of machine learning models, MLlib also provides a lot of utilities related to machine learning. MLlib supports different data formats for input and output, such as LibSVM, CSV, and PMML. To facilitate the training of machine learning models, MLlib has utilities of distributed linear algebra, statistical analysis, data preprocessing, and feature extraction.
- *Optimization Algorithms.* Different optimization algorithms are implemented in MLlib to train the provided machine learning models. Generalized linear models are trained by gradient optimization algorithms, including SGD [7], LBFGS [8–10], and OWLQN [11]. The training of decision trees borrows ideas from PLANET [12]. Recommendation models are optimized by alternating least squares.
- *Pipeline.* A real-world machine learning workload is not only about training. Typically, a machine learning job is a pipeline consisting of data preprocessing, feature extraction, model training, and model evaluation. MLlib naturally supports machine learning pipelines due to the data-flow capability of Spark. It is easy for users to develop machine learning pipelines since MLlib offers plenty of utilities.

Execution of MLlib Distributed training of machine learning models in MLlib relies on the master-slave architecture of Spark. There exist different distributed training schemes using MLlib, as shown in Fig. 4.8.

- *Broadcast data and model.* The master node (driver) loads the training dataset and initializes the model parameters. Then, the data and model parameters are broadcast to all the workers (executors). At each iteration, each worker performs computation (e.g., compute gradients) and updates the local model parameters. The master node collects the local model parameters from all the workers and merges them to obtain the new model parameters. The next iteration starts after the master again broadcasts the data and new model parameters. This distributed training scheme is only suitable for a small dataset and a small model due to repeated communication.
- *Broadcast model.* The worker nodes load their responsible data, and the master node broadcasts the initial model parameters to the worker nodes. At each iteration, each worker node performs the calculation and updates the local model parameters. The master node collects the local model parameters from the worker nodes and broadcasts the new model parameters again. This scheme can scale to a large-scale dataset. Nevertheless, the master node may become the system bottleneck for a large model.
- *Broadcast nothing.* Another distributed training scheme does not need the master node to broadcast and collect data. Each worker loads its responsible data and initializes the model parameters with the same strategy. At each iteration, each worker node computes statistics and updates the local model parameters. Afterward, the worker nodes use the *join* operator of Spark to share their local model parameters. This training scheme achieves better scalability than the above two approaches.

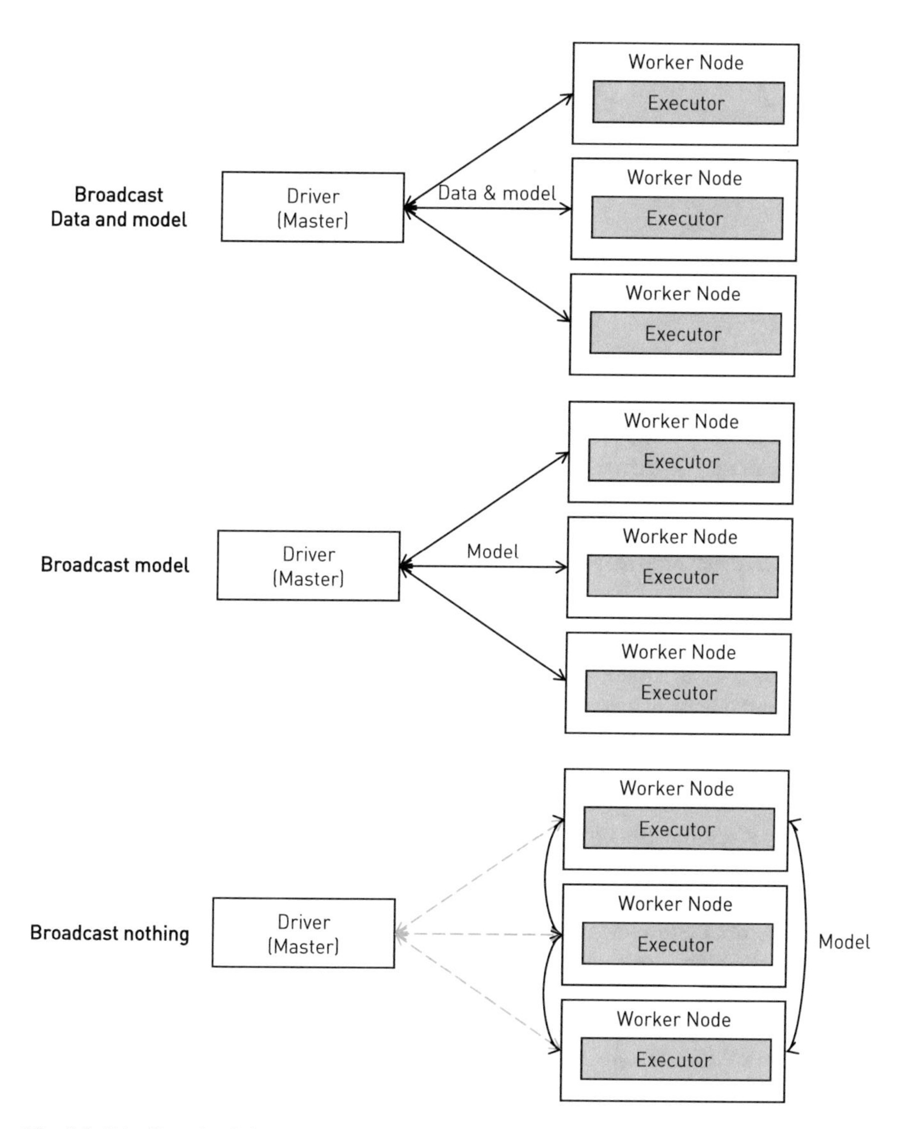

Fig. 4.8 Distributed training schemes of MLlib

4.1.2 Parameter Server Systems

Although MapReduce-based machine learning systems can handle diverse machine learning models due to the generality of MapReduce abstraction, they are unfit for many workloads in which the models are very large. If the model parameters are aggregated on the master node, the master becomes the bottleneck when the model is large. Even if adopting the *broadcast nothing* strategy in Fig. 4.8, the *join* operator still requires a single node to perform the aggregation.

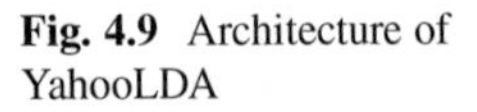
Fig. 4.9 Architecture of YahooLDA

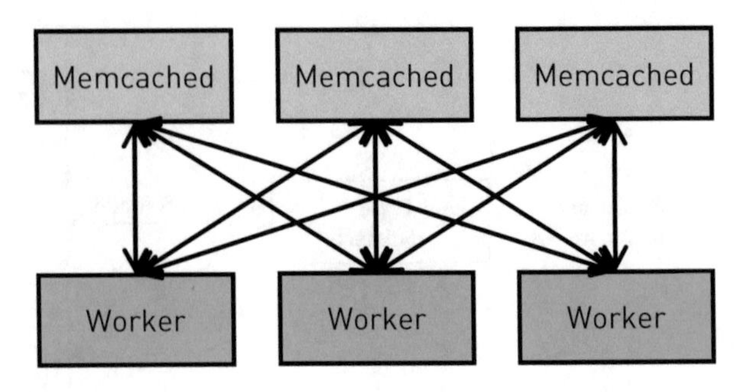

To address the problem of MapReduce-based systems for large models, a new framework called parameter server is proposed. Unlike the MapReduce framework, the parameter server framework establishes several machines together to aggregate the model parameters and thereby avoid the whole system being stalled by a single node. In this section, we present several existing parameter server systems designed for large-scale machine learning.

YahooLDA

Researchers from Yahoo [13] study the distributed training of large-scale LDA [14, 15]. LDA is a topic model in natural language processing. The input of LDA is a set of documents of words, and LDA assumes that each document is a mixture of k (generally a small number) topics and that each word's presence is attributable to one of the document's topics. While LDA is widely applied in both academia and industry, the training of LDA models often suffers scalability issues since the vocabulary size of documents can be extremely large.

To efficiently train LDA models, YahooLDA is designed by these researchers. Instead of aggregating the model parameters with one single machine, YahooLDA parallelizes the communication and storage of model aggregation through a distributed key-value store called *memcached*. As shown in Fig. 4.9, YahooLDA chooses to store a global state of the model parameters in several *memcached* nodes. If there are n workers and n servers, the communication cost and memory cost of one server is reduced by $n\times$ compared with using one sever. Since the global state is partitioned, it is required to periodically synchronize between the workers and server so that all the workers have identical local states as the global state on the servers. When this architecture is used to train machine learning models using gradient optimization algorithms, each worker performs local computation and updates the local state of model parameters. The local state and the global state of the model parameters are aggregated via an average operation.

Memcached provides many ready-to-use interfaces for users to manipulate the stored key-value data; therefore YahooLDA does not need to write the server code. However, the disadvantage of YahooLDA is the high latency of Memcached, especially when the workers and servers are located on different racks in a data center. In a later work published by the Yahoo researcher [13], they implement servers with common machines and split the global state of model parameters across

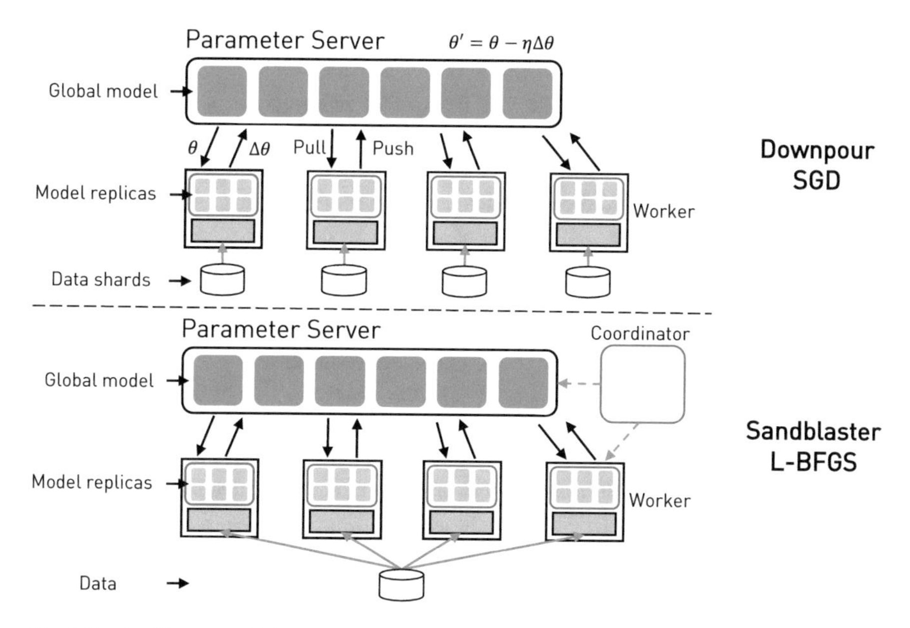

Fig. 4.10 Architecture of DistBelief

the servers. The workers and servers are colocated, that is, each machine runs a worker and a server simultaneously. Initially, the local states and global states of the model parameters are set to the same values. Each worker does local computation, updates the local model state. Once a synchronization point is reached, the worker estimates the change of the model state since the previous communication and transmits the change of model parameters to the servers. The servers apply the changes proposed by the workers to the global state. This new architecture reduces the latency to a large extent since the remote call to the server is nonblocking.

DistBelief

Dean et al. first introduce the concept of parameter server in their design system DistBelief [16]. To handle large models, DistBelief partitions the model across several machines (called parameter servers), as shown in Fig. 4.10. DistBelief automatically parallelizes the computation using multiple cores and manages the communication and synchronization during the training and inference.

DistBelief mainly focuses on neural networks and provides different levels of parallelism across the worker nodes. Specifically, DistBelief presents two large-scale gradient optimization algorithms—*Downpour SGD* and *Sandblaster L-BFGS*.

Downpour SGD Downpour SGD is a variant of distributed stochastic gradient descent algorithm with asynchrony. The training paradigm is presented below:

1. The training dataset is partitioned into several data shards, each of which is assigned to one worker node. A global state of the model parameter is also

partitioned into a set of model shards, and each shard is managed by one server node.
2. Each worker has a local replica of the model parameters and runs stochastic gradient descent over the responsible data shard.
3. At each iteration, each worker first asks the parameter server for the latest state of the model parameter and then updates the local model replica with the computed gradients.
4. The local updates of the model parameters are sent to the parameter server. Since the model parameters are partitioned across the servers, one worker needs to partition the local update using the same strategy as the parameter server and sends each shard of the update to the corresponding server.
5. Each parameter server applies the received update to the global model shard. Typically, the update is added to the model shard, possibly using a scaling factor such as the learning rate.
6. Once the current iteration is finished, each worker continues the processing of the next iteration.

Note that the data sent to the parameter server can also be gradients rather than the model updates. Downpour SGD introduces asynchrony for both workers and servers. The model shards on different parameter servers run asynchronously, and the workers run and update the local model replica asynchronously. The asynchronous nature of Downpour SGD makes it more tolerant to failures than standard SGD using the strict bulk synchronous protocol. If one worker node encounters failures, the other workers can keep running without being hampered by the failed worker. The downside of Downpour SGD is that the model shards on the parameter server may not receive the same number of updates in the same order at a certain moment. Moreover, the workers may not hold the same state of the local model replica. Therefore, there exist two kinds of inconsistencies in system—(1) between global model shards on the parameter server and (2) between local model replicas on the workers. To alleviate the model inconsistency, the authors propose to use the AdaGrad [17] technique to adaptively adjust the learning rate for each dimension of the model parameters.

Sandblaster L-BFGS As illustrated in Fig. 4.10, DistBelief designs Sandblaster, a batch gradient optimization framework, and implements L-BFGS-based on Sandblaster. Different from Downpour SGD, Sandblaster introduces more delicate manipulation of parameter storage and computation. A coordinator node is launched to assign tasks to the workers and servers. For example, the coordinator can ask one server to perform operations on its model shard, such as dot product, scaling, elementwise addition, and multiplication.

In Sandblaster L-BFGS, every worker has a local model replica and can access the whole training dataset. A global state of the model parameters is partitioned over the servers, and each server stores a model shard. The history gradients are also stored on the parameter servers so that the model update can be performed on the servers.

To address the straggler problem in the traditional implementations of L-BFGS, Sandblaster introduces a load-balancing strategy. During the processing of each batch, the coordinator assigns a small portion of the training job to each worker and successively assigns more portions when the worker is free. In this manner, the faster workers do more work than the slower ones, making the distributed processing more balanced and preventing the whole system is stalled by the stragglers. Furthermore, Sandblaster uses a "backup" strategy for the coordinator that assigns the same work to multiple workers and uses the first returned result. Compared with Downpour SGD in which the model synchronization between the workers and servers is frequent, the communication overhead during model synchronization in Sandblaster is much less because the workers only obtain the global state of model parameters at the beginning of each batch and only send gradients every several portions of the training work.

Petuum

Xing et al. propose a machine learning platform, called Pettum [18], based on the parameter server architecture. The architecture of Petuum is similar to the above parameter server systems. There are several parameter servers that establish distributed shared-memory storage for the workers. A scheduler is responsible for performing fine-grained control over the parallel computation. As an improvement over YahooLDA and DistBelief, Petuum provides carefully designed programming interfaces and eases the development of data- and model-parallel machine learning training.

Programming Interfaces To implement a program with Petuum, the user needs to write three functions shown in Algorithm 15—a central scheduler function `schedule()`, a parallel update function `push()`, and a central aggregation function `pull()`. A data-parallel program can be implemented by writing the push() function, while a model-parallel program can be implemented by writing all three functions schedule(), push(), and pull(). A PS object, which represents a model parameter stored on the parameter servers, can be manipulated from anywhere in the program. The PS object offers three functions for the users—`PS.get()` to obtain a parameter from PS, `PS.inc()` to add an update to the parameter, and `PS.put()` to write a new value to the parameter.

Architecture of Petuum Figure 4.11 shows the architecture of Petuum. There are three roles in Petuum—parameter server, worker, and scheduler. The functionality and implementation of each role are described below:

- *Parameter Server*: The parameter server provides access for users to read and write the model parameters. The servers store model shards in memory and expose distributed shared-memory interfaces for users to manipulate the model shards. As introduced above, there are three functions—`PS.get()`, `PS.inc()`, and `PS.put()`. The parameter server supports different kinds of synchronization protocols—BSP [19], SSP [20], and ESSP [21].

Algorithm 15 An example of Petuum program

x: features of training samples, y: desired outputs of training samples
$f(x, y; \theta)$: objective function, θ: model parameter (coefficients), θ^0: initial model parameter
W: # distributed workers, T: # iterations, b: batch size, η: learning rate

```
1: // the schedule function executed on the scheduler
2: function schedule () {
3:    A_local = PS.get(A)  // read parameter from PS
4:    PS.inc(A, Δ)  // add update to parameter
5:    svars = my_scheduling(DATA,A_local)  // choose variables for push() and return
6: }
7:
8: // the update function executed on W workers
9: function push (w = worker_id, svars = schedule ()) {
10:    A_local = PS.get(A)  // read parameter from PS
11:    // perform computation, send values to pull() or write to PS
12:    Δ1 = update_1(DATA,w,A_local)
13:    Δ2 = update_2(DATA,w,A_local)
14:    PS.inc(A,Δ1)
15:    return Δ2
16: }
17:
18: // the aggregation function executed on the scheduler
19: function pull (svars = schedule (), updates = (push(1), push(2), ..., push(W))) {
20:    A_local = PS.get(A)  // read parameter from PS
21:    new_A = aggregate(A_local, updates)
22:    PS.put(A, new_A)  // overwrite parameter
23: }
```

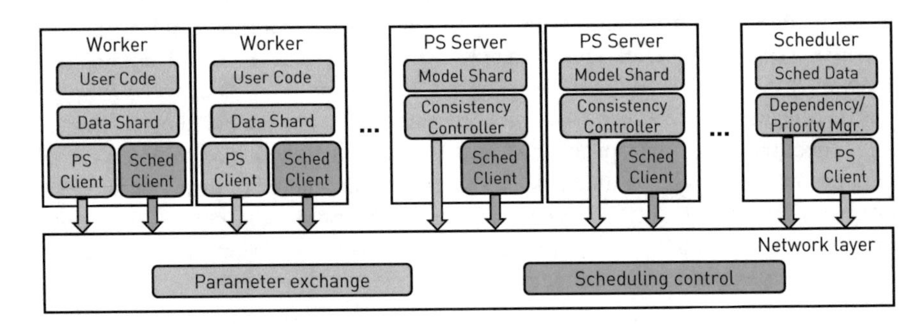

Fig. 4.11 Architecture of Petuum

- *Worker.* The processing procedure of each worker is as follows:
  1. The training dataset is partitioned over the workers. Each worker can access its data shard from a disk or distributed file system such as HDFS.
  2. At each training round, the worker obtains the scheduling result from `schedule()` and gets the responsible model parameters. The worker next runs `push()` over its local data shard, during which the worker can

manipulate the model parameters on PS using the interfaces of PS model. The scheduler may use the new model state to decide future scheduling.

- *Scheduler.* Petuum achieves model-parallel training using the scheduler. The scheduler allows users to control how the model parameters are updated. Specifically, the `schedule()` function outputs the portion of model parameters for each worker, and then the scheduler sends the identities of these parameters to the workers through the network. Petuum provides several ready-to-use schedules for users—`schedule_fix()` orders the model parameters in a static way (e.g., round-robin schedule), `schedule_dep()` allows the reordering of parameter updates by analyzing the dependency structure of the model parameters, and `schedule_pri()` allows prioritizing different subsets of model parameters according to the specific requirement of algorithm.

ParameterServer

Li et al. [22] engineer a parameter server system for scalable distributed machine learning, called ParameterServer. Compared with the aforementioned systems, ParameterServer makes contributions to efficient communication, flexible consistency, elastic scaling, and fault tolerance.

Architecture of ParameterServer The architecture of ParameterServer, shown in Fig. 4.12, contains a resource manager, a server group, and several worker groups.

- *Resource Manager.* The resource manager is responsible for allocating resources to all the running nodes.
- *Server Group.* The server group consists of several server nodes and a server manager. Each server node stores a partition of the global model parameters. The server nodes can communicate with each other to replicate and migrate the global

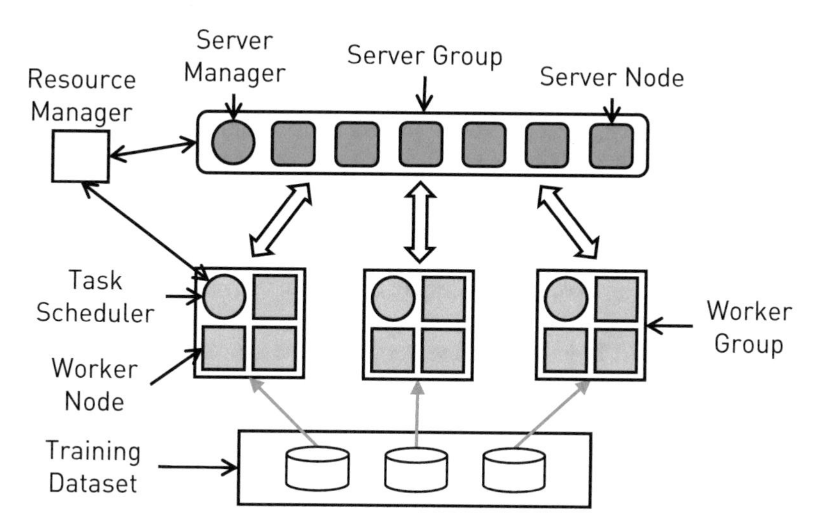

Fig. 4.12 Architecture of ParameterServer

model parameters. The server manager stores the metadata of the server nodes, such as the health status and the partition layout of model parameters.

- *Worker Group.* Each worker group contains several worker nodes and a task scheduler. Each worker node stores a partition of the whole training dataset, computes statistics such as gradients, pushes statistics to the server nodes, and pulls the latest model parameters from the server nodes. The task scheduler node assigns tasks to the worker nodes in the group. Due to the isolation of different worker groups, each worker group can train a different subset of the model parameters from other groups.

Management of Model Parameters The shared model parameters are represented as a set of (key,value) pairs and can be read and written locally and remotely using the keys. The value type of each pair is generally vector.

Each server node supports range push and pull operations. Assume R is a key range, the push function $\theta.push(R, dest)$ for a model θ sends the elements of θ within the range to the destination, either a server node or a server group. Likewise, the pull function $\theta.pull(R, dest)$ obtains the elements of θ within the range from the destination.

On the server node, the manipulation of model parameters can be customized by users. In addition to simple operations such as addition and multiplication, the server nodes can run user-defined aggregation functions to support complex updating mechanism.

Task Dependency and Consistency ParameterServer records the dependencies between tasks as the training job may contain many relevant subtasks. One task can be independent of another task or depend on the output of other tasks. Task dependencies are important for the implementation of consistency models. ParameterServer provides three different consistency models: (1) Sequential consistency: all tasks are executed sequentially, (2) Eventual consistency: all tasks run independently, and (3) Bounded delay consistency: the maximal delay between tasks is bounded by a threshold.

Management of Servers The parameter servers partition the keys of model parameters using consistent hashing [23, 24]. The keys and identities of server nodes are inserted into a hash ring. Following a counterclockwise direction, each server is responsible for the key range from its insertion point to the next point of other servers, called the master of this key range. To achieve load balancing and replication, a physical server is represented by several "virtual" servers in the hash ring.

When a new server is added to the server group, the identity of the new server is inserted into the hash ring. The server manager assigns the new server as the master for its responsible key range and assigns it as the slave for several additional ranges. Then the server manager broadcasts the change of hash ring, based on which other server nodes update their key ranges. The departure of an existing server node is handled in a similar manner as server join.

Management of Workers When a new worker is added to a worker group, the task scheduler assigns a portion of the training data to the new worker. The new worker loads its data shard from the data source, and other affected workers may need to drop overlapping data. When an existing worker is lost, ParameterServer provides two options for users—either start a new worker as a replacement or continue without replacing the departure worker. For some workloads in which losing a small portion of the training data barely affects the model quality, it is often not necessary to spend time launching a new worker.

Angel

Angel is an open-source large-scale machine learning system developed by researchers from Tencent Inc [25, 26]. Angel mainly aims at training machine learning models over parameter server architecture in real industrial environments. To this end, Angel proposes a series of optimization w.r.t. communication and integrates itself with the ecosystem of big data processing.

Architecture of Angel As shown in Fig. 4.13, there are four roles in the architecture of Angel—client, master, worker, and server.

- *Client.* The client is the entry of an Angel application, which loads the user programs and submits the application to the resource manager Yarn. The client keeps running until the application finishes or fails.

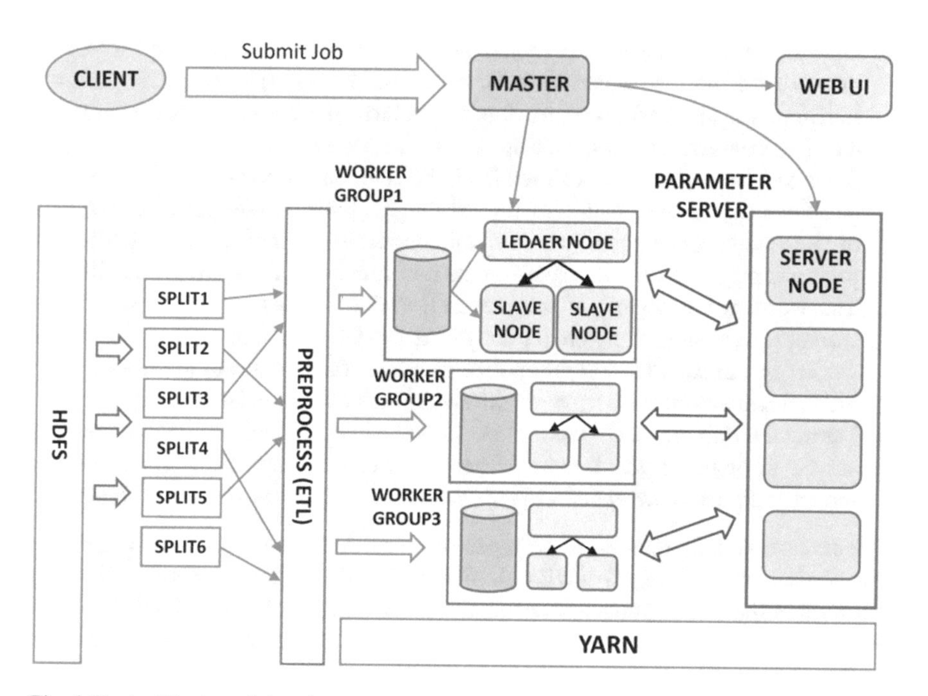

Fig. 4.13 Architecture of Angel

- *Master.* Master is launched to manage the execution of an application. including requesting resources from the resource manager, launching workers and servers, and terminating them. Master also handles the synchronization between workers and servers.
- *Worker.* Workers execute the program written by users. Each worker loads data from a distributed file system such as HDFS, reads model parameters from the parameter servers, computes statistics, and sends updates to the servers.
- *Server.* The servers together store model parameters, and each server stores a model shard. The servers provide distributed in-memory access for the workers to manipulate the model parameters.

Lifecycle of Angel The execution of an Angel application generally contains the following steps:

1. A user submits the application code to Yarn through the client. Yarn first launches the master to manage the execution of the application. Then, the master requests resources from Yarn to allocate containers for workers and servers. The number of workers and servers, along with their required resources, can be configured by the user.
2. After the workers and servers are launched, the servers initialize the model parameters according to the configurations provided by the user. First, the metadata of the model parameter, including the model name, model size, and partition locations, is parsed and stored in the master. Then, each server gets the metadata of responsible model partitions from the master and allocates memory for the model partitions. The servers provide `get` and `increment` functions for the workers—`get` function returns the current model parameters and `increment` function adds updates to the model parameters.
3. Since the training dataset is stored in HDFS, the master obtains the location and partitions the training data into several ranges. Each worker gets the metadata of its responsible data partition from the master, loads the data from HDFS, and performs preprocessing to transform the raw data into a compatible data format.
4. The workers are organized into several worker groups by the master. Each worker group has a complete replica of the model parameters. There is a leader worker inside each worker group that schedules the updates to the model. This hierarchical architecture supports different kinds of parallelism strategies.
5. During the training, each worker first pulls the latest model parameters from the servers through the `get` function. Then, the worker computes the updates to the model and pushes the updates to the servers through the `increment` function.

Parallelism Angel supports three types of parallelism—*data parallelism*, *model parallelism*, and *hybrid parallelism*. In data parallelism, each worker is assigned a partition of the training data and maintains a local replica of the model parameters. Angel implements data parallelism by launching only one worker in each worker group. Different from data parallelism, model parallelism does not partition the training dataset. Instead, the model parameters are partitioned over the worker, and each worker holds a partition of the whole model. Since each worker runs independently on a different subset of the model, a coordinator is needed to schedule the updates of the model. Angel implements model parallelism by organizing all the workers into one worker group and assigning the leader worker as the coordinator. Hybrid parallelism can be implemented by utilizing the hierarchical structure of worker groups and workers. The training dataset is partitioned over the worker groups, and each worker group maintains a complete replica of the model parameters. Inside one worker group, one worker is selected as the leader, and other workers are considered as slaves. Each worker is responsible for a partition of the model, and the model updates generated by them are scheduled by the leader.

Management of Training Data The training dataset is a 2D data matrix where each row is a training instance and each column represents a feature. Angel offers three strategies to partition the training dataset over the workers.

- *Replication.* The whole training dataset matrix is replicated on every worker.
- *Horizontal partitioning.* The training data matrix is horizontally partitioned across the workers so that each worker holds a subset of rows (instances).
- *Vertical partitioning.* For some machine learning models that can be parallelized between different features, Angel can vertically partition the data matrix so that each worker stores several columns of the matrix. In this case, all the values of a certain feature are at a single worker.

Management of Model Parameters Similar to the management of training data, Angel abstracts model parameters in the format of matrices, such as the weight vector in linear models and word-topic count matrix in LDA [14]. Angel partitions the model parameters with 2D partitioning (both horizontally and vertically) into rectangle blocks. The number of blocks, i.e., row size and column size, can be configured by users. Each block is the communication unit between servers and workers. To achieve higher parallelism, the number of blocks is normally larger than the number of servers.

Synchronization Angel supports various kinds of synchronization protocols, including bulk synchronous protocol [19], asynchronous protocol [16], and stale synchronous protocol [20]. Before the execution of one iteration, each worker can obtain the minimum iteration of all the workers from the master, based on which the worker determines whether it is allowed to run the next iteration.

Pipeline Support Angel is closely integrated with Spark. It can benefit from the utilities provided by Spark, including data loader, data preprocessing, feature engineering, linear algebra library. Besides, Angel can be assembled into Spark pipelines as it is abstracted as a module of Spark. For example, a Spark pipeline can be created by (1) first preprocessing the training dataset using naive Spark, (2) training a large-scale machine learning model with Angel, and (3) evaluating the fitted model using Spark. Pipeline processing by combining Angel and Spark is able to avoid frequent writing data to the disk between connected modules.

Supported Machine Learning Models Angel provides many popular machine learning models that can be easily used by practitioners, including logistic regression,

Algorithm 16 Exact greedy splitting algorithm

D: training dataset, M: # features, N: # instances, K: # split candidates
g_i, h_i: first-order and second-order gradients of an instance

1: $gain_{max} = 0, G = \sum_{i=1}^{N} g_i, H = \sum_{i=1}^{N} h_i$
2: **for** $m = 1$ to M **do**
3: $G_L = 0, H_L = 0$
4: **for** j in $sorted(D, by\ feature\ m)$ **do**
5: $G_L = G_L + g_j, H_L = H_L + H_j$
6: $G_R = G - g_j, H_R = H - H_j$
7: $gain_{max} = max(gain_{max}, \frac{G_L^2}{H_L+\lambda} + \frac{G_R^2}{H_R+\lambda} - \frac{G^2}{H+\lambda})$
8: **end for**
9: **end for**
10: Output the split with max gain

support vector machine, linear regression, LDA, gradient boosting decision tree (GBDT), and Matrix Factorization.

Deep Learning Support The latest version of Angel has supported many deep learning models. Angel embeds PyTorch [27], a popular deep learning library, in their worker [28]. Each worker runs PyTorch in C runtime and communicates with the servers to manipulate model parameters. In this manner, Angel supports different kinds of deep learning models, such as convolutional neural networks (CNNs) and graph neural networks (GNNs), and accelerates the execution of large-scale models using the parameter server architecture.

4.2 Specialized Machine Learning Systems

General machine learning systems support a wide range of machine learning models while lacking specific optimization for each model individually. We find that these systems mostly focus on gradient boosting decision tree (GBDT). The necessity of building specialized systems for tree models derives from the unique training paradigm of GBDT. Although the training of GBDT also uses gradients, it is different from the models trained by gradient descent algorithms. We present two widely used systems in this section—XGBoost and LightGBM. The concepts, principles, and training methods of GBDT are elaborated in Sect. 3.3.

XGBoost

Chen and Guestrin propose a scalable training system for gradient boosting decision tree, called XGBoost [29].

Architecture of XGBoost In XGBoost, the workers use MPI to communicate with each other. Since the users can write user-defined MPI programs, the complex training scheme and a series of specific optimizations.

Split Finding Strategy The key technique in training GBDT is to find the best split of each tree node. XGBoost supports both exact greedy and approximate algorithms for split finding.

- *Exact greedy algorithm.* The exact greedy algorithm enumerates all the possible splits for all the features. The procedure of exact splitting is presented in Algorithm 16. XGBoost sorts all the values of each feature and chooses all the possible splits by looping the sorted values, calculates the split gain based on gradient statistics and selects the optimal split with the maximum split gain. The disadvantage of the exact greedy algorithm is that it is inefficient for large-scale data.
- *Approximate algorithm.* To avoid enumerating all possible splits, an alternative is to employ an approximate approach, as described in Algorithm 14. XGBoost first build a quantile sketch [30, 31] to summarize the distribution for each feature.[1] Then, a fixed number of splits are generated from the quantile sketch according to the percentiles of feature distribution. Afterward, XGBoost uses these splits to aggregate the features into buckets, calculate gradient statistics and find the best split using the same accumulation as the exact greedy algorithm. XGBoost provides two ways to generate quantile splits—(1) the global strategy proposes the candidate splits before the building of trees and uses these splits for all tree nodes; (2) the local strategy updates the candidate splits after splitting each tree node. The global proposal usually needs more candidate points because candidates are not refined after each split, while the local proposal refines the candidates after splits, which is more appropriate for deeper trees.

  XGBoost further proposes a weighted quantile sketch that is more suitable for GBDT. Recall the loss function of GBDT, which can be rewritten as:

$$\begin{aligned} F^{(t)} &\approx \sum_{i=1}^{N} \left[l(y_i, \hat{y}_i^{(t-1)}) + g_i f_t(x_i) + \frac{1}{2} h_i f_t^2(x_i) \right] + \Omega(f_t) \\ &= \sum_{i=1}^{N} \frac{1}{2} h_i \left(f_t(x_i) + \frac{g_i}{h_i} \right)^2 + \Omega(f_t) + constant \end{aligned} \tag{4.1}$$

  The loss function can be considered weighted squared loss. As there is no quantile sketch designed for this weighted setting, the authors design a weighted variant of quantile sketch [29].

Sparse-Aware Strategy Since the datasets in many real-world workloads are sparse, XGBoost proposes a sparse-aware strategy to handle sparse cases including missing values, zero values, or one-hot encoding features. Briefly speaking, XGBoost proposes adding a default direction in each tree node. When a value is missing in a sparse matrix, the instance is classified into the default direction. After reading

[1] https://datasketches.github.io/.

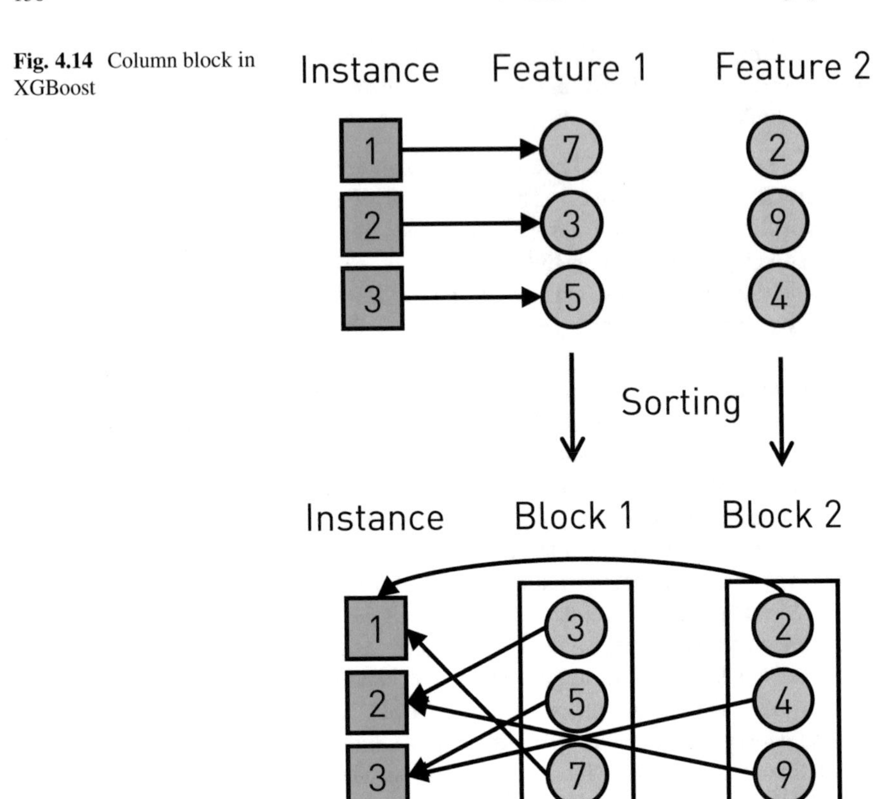

Fig. 4.14 Column block in XGBoost

all the values, the sparse-aware algorithm tries two options for the missing values, either putting the missing values to the left or the right. Evaluations over the Allstate-10K dataset show that the sparsity-aware algorithm achieves a 50× speedup compared with the naive implementation.

Column Block During the training of a GBDT model, the process of split finding needs to frequently sort the training data for each feature. XGBoost designs a block structure to reduce the cost spent on sorting. The data in each block is stored in the format of CSC (compressed sparse column), and each column (feature) is sorted by the corresponding feature value. As shown in Fig. 4.14, XGBoost sorts the values of each feature, having a pointer to the corresponding instance for each feature value. XGBoost creates all the blocks before training and reuses them during the whole training process.

Other Optimizations XGBoost uses several other optimizations for distributed training.

- *Shrinkage.* The shrinkage trick is adopted by giving a scaling factor $\eta < 1$ for the output of every tree. Shrinkage reduces the contribution of each tree, and hence prevent overfitting.

- *Column subsampling.* Another technique used to prevent overfitting is column subsampling, also called feature subsampling. By choosing different subsets of features for the trees, the diversity of the built trees is enhanced. Besides, feature subsampling can accelerate the training speed.
- *Out-of-core computation.* When the volume of the dataset is too large, the memory is not enough to store all the blocks. XGBoost supports out-of-core computation by dividing the dataset into several blocks and storing some of the blocks on disk. To alleviate slow disk access, XGBoost prefetches the block into a memory buffer so that the data loading and computation overlap.

LightGBM

Although XGBoost has achieved great successes in both academia and industry, it reveals performance issues for high-dimensional and large-scale datasets. The reasons are two-fold. First, the traditional method needs to scan all the training instances to estimate the gain brought by each candidate split. The computation cost can be too expensive for large-scale datasets. Second, distributed training of GBDT needs to aggregate the gradient statistics, i.e., gradient histograms. The size of the gradient histograms is affected by the number of features. For high-dimensional datasets, the gradient histograms that need to be aggregated can be very large, making the communication become the bottleneck. LightGBM [32], a distributed framework for gradient boosting decision tree,[2] is proposed to efficiently train GBDT in distributed environments. LightGBM decreases computation and communication in two ways—reduce the number of training instances or the number of used features.

Architecture of LightGBM Similar to XGBoost, LightGBM also adopts MPI to implement communications among distributed nodes.

Reducing the Number of Training Instances LightGBM proposes a novel sampling method, called gradient-based one-side sampling (GOSS), that can reduce the number of training instances and achieve comparable accuracy. Unlike the weighted sampling method used in algorithms such as AdaBoost, there are no explicit sample weights in GBDT. Therefore, a weight-based sampling method cannot be applied.

Considering the main statistics used in GBDT are gradients, LightGBM tries to leverage gradients to perform sampling. Intuitively, if the gradient of an instance is small, this instance has a small impact on the model as its training error is small. However, directly discarding these instances with small gradients changes the data distribution and may produce a less accurate model. To avoid this situation, GOSS keeps the instances with large gradients and randomly samples the instances with small gradients. To compensate for the negative impact on data distribution, GOSS uses a constant value for small gradients when computing the information gain. In the implementation of GOSS, LightGBM first sorts all the training instances by the absolute values of their gradients and then selects the top $a\%$ training instances.

---

[2] https://github.com/microsoft/LightGBM.

Afterward, GOSS samples $b\%$ of the training instances from the remaining $1 - a\%$ instances. When computing the information gain, GOSS considers the discarded instances by multiplying a constant $\frac{1-a}{b}$.

Reducing the Number of Features For datasets with high-dimensional features, LightGBM proposes a method called exclusive feature bundling (EFB) to reduce the number of features. The basic idea is that many features in the sparse feature space are mutually exclusive, meaning that they never take nonzero values simultaneously. Therefore, these features can be bundled into a single new feature. The authors prove that the built gradient histograms remain the same after transforming the original features into bundled features, while significantly decreasing the computation complexity without hurting the model accuracy.

Other Optimizations In another relevant paper of LightGBM [33], the authors propose a new algorithm, called parallel voting decision tree (PV-Tree), to reduce the communication cost. PV-Tree partitions the training dataset over the workers and performs the split finding of tree node using a two-phase strategy—a local voting phase and a global voting phase. At each iteration, the local voting phase makes each worker select the top-k features w.r.t information gain on its local training data, and the global voting phase uses majority voting to select the top-$2k$ features. The gradient histograms of these $2k$ features are then aggregated from the workers to calculate the optimal split, including the split feature and the split value. PV-Tree only consumes a very small communication cost with a theoretical guarantee to find a near-optimal model.

4.3 Deep Learning Systems

Recent decades have witnessed the blooming of deep learning. The driving force of the huge success of deep learning derives from the exponentially increasing big data and the advancement of systematic processing capabilities over new hardware such as GPUs. Many systems are designed specifically for training deep learning models efficiently, such as Caffe, TensorFlow, PyTorch, and MXNet. We do not take some CPU-targeted systems, such as Theano, because GPU has been the mainstream hardware for deep learning models.

Caffe

Jia et al. propose Caffe, a framework for deep learning models on GPUs [34]. Caffe provides user-friendly programming interfaces, which are separated from the actual execution. Caffe may be the first-ever widely adopted deep learning system over GPUs.

Data Management Caffe stores data in the format of 4-dimensional arrays called *blobs*. Blobs provide in-memory accessing interfaces to manipulate the data (e.g., images) and model parameters. Typically, a batch of training data is loaded from the disk to CPU as a blob, moved to the GPU device from the CPU host, and used to perform GPU computations. After the training is finished, the model is saved to the disk using Google Protocol Buffers.[3]

Layer Abstraction Caffe abstracts a neural network as the combination of layers. Each Caffe layer takes blobs as input and outputs blobs for other layers. There are two key functionalities in one Caffe layer—*forward* and *backward*. The *forward* pass calculates the outputs of one layer and feeds to the subsequent layers. The *backward* pass takes the gradients of its outputs from the subsequent layers, calculates the gradients w.r.t. the corresponding model parameters and its inputs, and then back-propagates the gradients to previous layers.

Caffe provides many popular layers for users, including convolutional layer, pooling layer, activation layer, normalization layer, SoftMax layer, etc. Users can also customize their own layers following the programming pattern of Caffe.

Execution After creating a neural network by Caffe, the computation of the network often starts with a data layer and ends with a loss layer. As shown in Fig. 4.15, the training dataset is fed into the data layer as minibatches sequentially. The network generates the outputs (stored as blobs) of each layer using the *forward* pass. In particular, the final loss layer takes predictions from its previous layers and the labels of the training data. Then, the layers yield the gradients using the *backward* pass. Caffe provides a standard stochastic descent algorithm with techniques such as decay learning rate and momentum. The execution of the Caffe network can be run on CPUs or GPUs, and the switch between CPUs and GPUs is seamless to the users.

Implementation Caffe is developed with C++ and CUDA (the GPU computation library provided by Nvidia). Besides, it has close integration with Python, NumPy, and MATLAB.

TensorFlow

Google researchers develop a system for large-scale deep learning called TensorFlow [35]. TensorFlow abstracts the computation of a deep learning model as a graph of dataflow. TensorFlow was originally designed for heterogeneous environments, such as CPUs, GPUs, ASICs, and TPUs. Due to the elegant abstraction and industry-targeted flexibility, TensorFlow is widely used in many research works and real-world applications.

Dataflow Graph Abstraction Previous deep learning systems, such as DistBelief and Caffe, abstract the computation of a deep learning model as the combination of layers (e.g., convolutional layer, pooling layer, and fully connected layer). However, this kind of abstraction is coarse-grained and inefficient for users to implement new network structures. TensorFlow proposes a dataflow abstraction that represents both the computations and the states in one neural network as a graph. In such one computation graph, the vertices represent different kinds of

[3] https://github.com/protocolbuffers/protobuf.

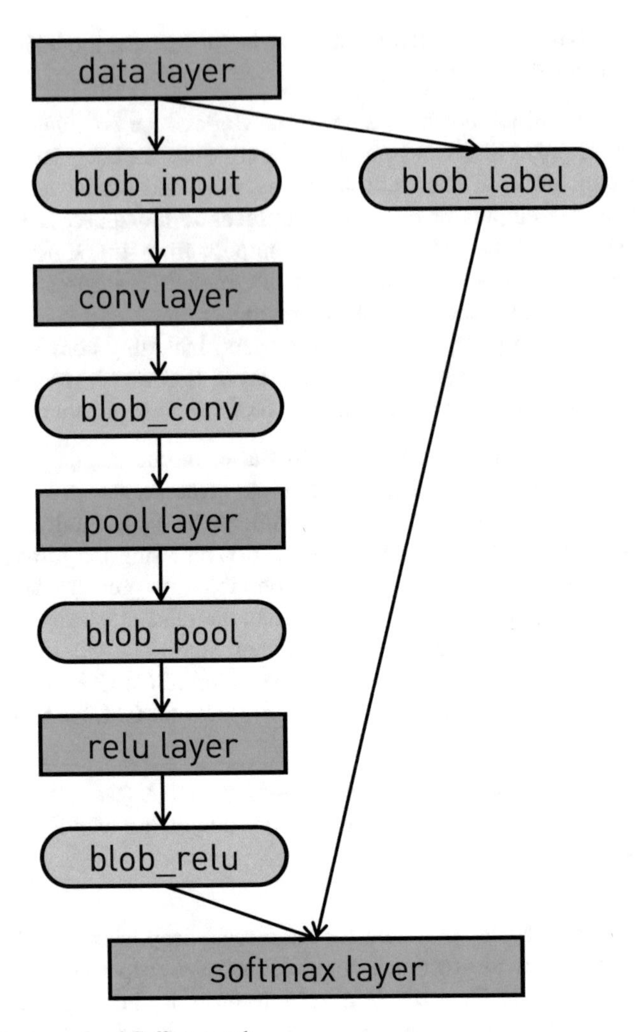

Fig. 4.15 An example of Caffe network

mathematical operators and the edges represent tensors (multidimensional arrays) between vertices. Furthermore, TensorFlow can also represent the mutable state and the operations updating them as vertices in the dataflow graph, allowing the possibility of different update rules.

The fine-grained abstraction of TensorFlow unifies the mathematical computation, the shared parameters (states) and state management (update rules) in a single programming model, making it easier and flexible for users to implement novel neural network structures and diverse parallel strategies. Especially, the mutable state support in the dataflow renders TensorFlow efficient for very large model parameters and enables different optimization algorithms and consistency protocols. This mechanism is similar to the parameter server architecture that provides in-memory interfaces to manipulate the shared model parameters.

Listing 4.1 An Example Script of TensorFlow for Training MNIST Dataset

```
# Placeholder for inputs and labels
x = tf.placeholder(tf.float32, [BATCH_SIZE, 784])
y = tf.placeholder(tf.float32, [BATCH_SIZE, 10])

# Define model parameters as variables
W_1 = tf.Variable(tf.random_uniform([784, 100]))
b_1 = tf.Variable(tf.zeros([100]))
W_2 = tf.Variable(tf.random_uniform([100, 10]))
b_2 = tf.Variable(tf.zeros([10]))

# Construct the model using dataflow graph abstraction
layer_1 = tf.nn.relu(tf.matmul(x, W_1) + b_2)
layer_2 = tf.matmul(layer_1, W_2) + b_2

# Define the loss and the optimization algorithm
loss = tf.nn.softmax_cross_entropy_with_logits(layer_2, y)
train_op = tf.train.AdagradOptimizer(0.01).minimize(loss)

# Train the model
with tf.Session() as sess:
  # Randomly initialize the model parameters.
  sess.run(tf.initialize_all_variables())
  for step in range(NUM_STEPS):
    # Load one mini-batch of input data.
    x_data, y_data = ...
    # Run one training round
    sess.run(train_op, {x: x_data, y: y_data})
```

Listing 4.1 shows a script for defining and training a deep learning model in TensorFlow. Placeholders are used to store the currently processed input training instances and labels, and the training instances are represented as tensors. The model parameters are defined as variables in TensorFlow. The trained neural network model is created by constructing a dataflow using TensorFlow *Operators*. After defining the loss function and the optimization algorithm, the model is trained iteratively by executing the dataflow graph over each minibatch of input data. Below we define these fundamental elements in the dataflow graph of TensorFlow.

- *Tensors.* TensorFlow stores all the data, e.g., the inputs and the outputs of mathematical operators, as tensors (multidimensional arrays). The supported data types include `int32`, `float32`, or `string`.
- *Operations.* A TensorFlow operation takes as input one or multiple tensors, and outputs one or multiple tensors. TensorFlow provisions various kinds of operations, such as `Const`, `Add`, and `MatMul`. Each operation can have various attributes that determine its functionality. For instance, `Const` takes no input and a single output of a compile-time constant value; and `Add` performs elementwise sum over two tensors of the same element type.
- *Variables.* TensorFlow enables shared stateful states through the variable operation. A variable operation has a mutable buffer that is used to store the shared

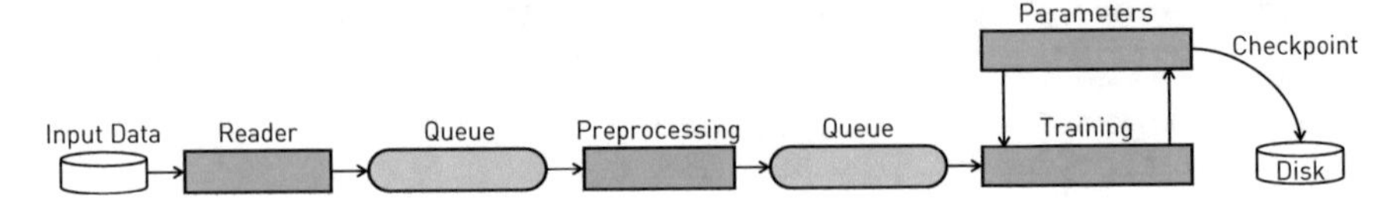

Fig. 4.16 A typical TensorFlow pipeline

model parameters. It has no input and gives a reference handle that can be utilized to read and write the buffer.

- *Queues.* To support concurrent access of tensors, TensorFlow has a queue style implementation. The `Queue` operation has a reference handle like `Variable` that can be consumed by standard queue operations such as `Eneueue` and `Dequeue`. For example, `FIFOQueue` allows a first-in-first-out accessing order for a queue of tensors.

Pipeline Execution Once a TensorFlow application is created, it is executed according to the defined dataflow graph. Figure 4.16 showcases a dataflow pipeline of TensorFlow. The pipeline consists of a reader subgraph that loads the input data, a preprocessing subgraph, a network subgraph needed to be trained, and a periodical checkpoint subgraph. In this pipeline, different subgraphs are processed concurrently, which interact with each other through shared variables and queues. The data reader subgraph loads the input data from a distributed file system into a queue. The preprocessing subgraph runs concurrently to transform the original data into the compatible format and fills the input queue for the training subgraph. In the training subgraph, multiple concurrent subgraphs take inputs from the queue and update the shared parameters. For the sake of fault tolerance, the periodical checkpoint subgraph runs independently to store the current state of shared parameters in a distributed file system.

As can be observed, the shared variables and queues enable partial and concurrent execution of a TensorFlow application. This flexible scheme also provides flexibilities for users to design new network structures, optimization algorithms, synchronization protocols, and parallel strategies.

Distributed Execution The dataflow abstraction of TensorFlow naturally supports distributed execution since the subgraphs can be deployed on different physical machines and the subgraphs can communicate with each other. Each operation in the computation graph is associated with a specific device, e.g., a CPU or a GPU. Each device executes a kernel for the assigned operation.

A challenge during distributed training is how to place the operations given a set of devices. The placement algorithm in TensorFlow chooses a feasible set of devices for each operation, selects the operations that must be colocated, and calculates a suitable device for each colocated operation group. The colocated operations are determined according to implicit or explicit constraints of the graph. For example, a stateful operation and its state must be placed on the same device. Besides, the users can configure their constraints on the placement of operations, or manually place each operation to a specific device.

Once the placement algorithm is finished, each device is assigned a subgraph. Each subgraph contains several operations, a `Send` operation, and a `Recv` operation. `Send` and `Recv` operations represent the edges between device boundaries. `Send` transmits the tensors on the device to other devices, while `Recv` receives tensors from other devices.

Dynamic Control Flow TensorFlow introduces dynamic control flow to support advanced deep learning models that have conditional and iterative control flow. Taking RNN as an example, RNN contains a recurrence relation, which iterates over sequences having variable lengths. Dynamic control flow supports these deep learning models by introducing conditional (if statement) and iterative (while loop) programming interfaces in the dataflow graph.

Dynamic control flow is achieved using the deferred execution principle of TensorFlow. The execution of a TensorFlow application contains two phases—the first phase defines the dataflow graph using placeholders for the input data and variables for the states; the second phase actually executes the dataflow graph on the chosen devices. Since the actual execution is deferred, the dynamic control flow is achieved accordingly.

Synchronization TensorFlow supports three types of synchronization protocols—asynchronous replication, synchronous replication, and synchronous with backup workers.

- *Asynchronous Training.* TensorFlow is originally designed for asynchronous training as SGD is tolerant to asynchrony. At the beginning of each iteration, each worker reads the current state of shared model parameters, calculates the gradients, and applies the gradients to the shared model parameters. Asynchronous training can assure high throughput in the presence of stragglers but may suffer unstable convergence as a result of stale model parameters.
- *Synchronous Training.* Other than asynchronous training, TensorFlow also supports traditional synchronous training. Some works have studied whether synchronous training is suitable for distributed learning [36]. Since many large-scale deep learning applications run on hundreds of GPUs, synchronous may be more suitable as many stale parameters influence the stable convergence. TensorFlow implements synchronous training using the queue mechanism. A blocking queue is used to ensure that all the workers read the same model parameters at the same barrier, and another queue is used to receive gradients from the workers and apply the accumulated gradients to the model parameters. Synchronous training guarantees stable convergence, while the downside of synchronous training is obvious—the overall throughput is limited by the stragglers.
- *Synchronous Training with Backup Worker.* To address the straggler problem in synchronous training, TensorFlow implements the backup strategy [1, 36]. TensorFlow launches several backup workers and lets them perform backup training when detecting stragglers. The aggregation queue takes the first N updates and updates the shared model parameters, where N is the number of workers.

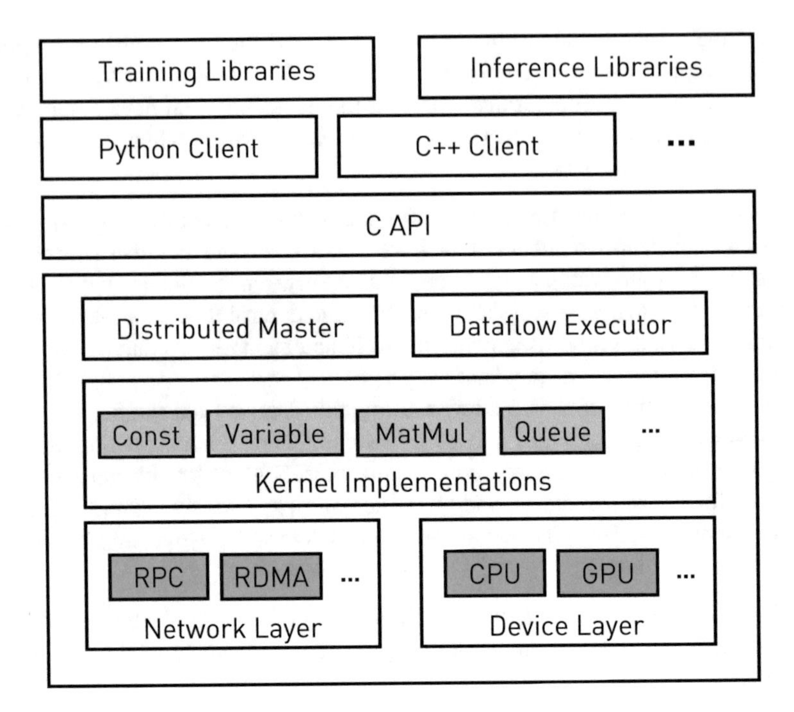

Fig. 4.17 Architecture of TensorFlow

Implementation Figure 4.17 shows the architecture of TensorFlow implementation. The core runtime is implemented with C and provides C APIs for high-level libraries. Beyond the basic C API, TensorFlow engineers more user-friendly APIs with other languages, such as Python and C++. The core libraries of TensorFlow (e.g., networking, devices, and kernels) can run on different kinds of operating systems, including Linux, Mac OS, Windows, etc. TensorFlow provides more than 200 operations, including mathematical operations, array manipulations, stateful operations, and control flow. These operations can run on both CPUs and GPUs.

The distributed master is responsible for translating user requests into execution plans. Specifically, it receives the dataflow graph and partitions the graph into subgraphs over available devices. Each dataflow executor gets commands from the master, and executes the kernels in its local subgraph. The chosen kernels are assigned to a local device by the dataflow executor and run in parallel.

PyTorch

PyTorch is proposed by researchers from Facebook [27] and provides imperative Python-style programming APIs and efficient performance over new hardware like GPUs.

Listing 4.2 Neural Network Structures Implemented with PyTorch.

```
class LinearLayer(Module):
  def __init__(self, in_size, out_size):
    super.__init__()
    t1 = torch.randn(in_size, out_sie)
    self.w = nn.Parameter(t1)
    t2 = torch.rand(out_size)
    self.b = nn.Parameter(t2)

  def forward(self, input):
    t = t torch.mm(input, self.w)
    return t + self.b

class MyNet(nn.Module):
  def __init__(self):
    super.__init__()
    self.conv = nn.Conv2d(1, 128, 3)
    self.fc = LinearLayer(128, 10)

  def forward(self, x):
    t1 = self.conv(x)
    t2 = nn.functional.relu(t1)
    t3 = self.fc(t2)
    return nn.functional.softmax(t3)
```

Pythonic Programming Model Since most, if not all, data scientists and practitioners prefer developing deep learning applications using the Python ecosystem, PyTorch introduces a graph metaprogramming based approach to preserve the imperative programming model of Python. Users can use the Python programming model to define all aspects of deep learning applications, including defining layers, defining networks, loading data, optimizing, and parallel processing.

This Pythonic programming model also makes it easier for users to implement new network structures. As shown in Listing 4.2, a neural network layer is defined as a Python class by inheriting the `Module` class in PyTorch. There are two major functions that need to be implemented—(1) the `__init__` function creates necessary parameters, and (2) the `forward` function processes the input activations. In addition to layers, an entire neural network can be implemented in a similar way. The `__init__` function defines the network structure by composing the PyTorch modules, such as 2D convolution, matrix multiplication, ReLU activation, dropout, SoftMax, and user-defined modules.

Listing 4.3 presents an example script of training neural networks using PyTorch. The training procedure contains the following steps:

1. Load the training dataset, transform the original data into the supported format, and create the data loader.
2. Construct the model structure and optimizer.
3. At each iteration, the training engine takes one minibatch of data from the data loader and sets the gradients to zero. The forward pass of the model calculates the

Listing 4.3 An Example Training Script of PyTorch.

```
# Specify the used device
device = torch.device("cuda" if use_cuda else "cpu")

# Create Tensors to hold input and outputs.
# Load the dataset and perform preprocessing
transform=transforms.Compose([
  transforms.ToTensor(),
  transforms.Normalize((0.1307,), (0.3081,))
  ])
dataset1 = datasets.MNIST('../data', train=True, download=True,
    transform=transform)
dataset2 = datasets.MNIST('../data', train=False,
    transform=transform)
train_loader = torch.utils.data.DataLoader(dataset1)
test_loader = torch.utils.data.DataLoader(dataset2)

# Construct the model by instantiating the class defined above
model = MyNet().to(device)

# Construct the optimizer
optimizer = torch.optim.SGD(model.parameters(), lr=1e-8,
    momentum=0.9)

for epoch in range(1, n_epochs + 1):
  for batch_idx, (data, target) in enumerate(train_loader):
    # Load a mini-batch data from the data loader
    data, target = data.to(device), target.to(device)
    # Set gradients to zero
    optimizer.zero_grad()

    # Forward pass: Compute prediction by passing input to the
        model
    output = model(data)

    # Compute the loss
    loss = F.nll_loss(output, target)

    # Perform the backward pass, and update the model
        parameters.
    loss.backward()
    optimizer.step()
```

prediction with the input data and the loss. Then, the backward pass is executed and the computed gradients are applied to the model parameters.

Automatic Differentiation Deep learning models are generally trained by back-propagation. In the above programming model of PyTorch, users do not need to explicitly implement the backward process. This is owing to the feature of PyTorch that it can automatically calculate the gradients according to the model structure defined by the user. PyTorch uses the operator overloading approach to implement automatic differentiation, which creates a representation of the function every time it is executed. Based on the overloading representations, PyTorch implements automatic differentiation, which computes the gradient of a scalar output with respect to a multivariate input.

Control and Data Flow As we have discussed in the introduction of TensorFlow, many deep learning models have control operations in the structure, e.g., if statement and loop. PyTorch runs the control flow on CPUs, while it is implemented by Python to assure sequential invocations of operators.

In contrast, PyTorch runs operators on GPU asynchronously using CUDA stream. This mechanism makes it possible to overlap the execution on CPUs and tensor operations on GPUs, bringing higher hardware utilization.

Distributed Execution If a machine learning model is trained in a distributed setting, the running workers need to communicate model parameters (in the format of tensors). PyTorch organizes the workers into a group in which each worker is assigned a rank. Point-to-point communication is achieved through the `send` and `recv` functions using the corresponding ranks of source and destination workers. Point-to-point communication provides flexibility for users to implement fine-grained and complex communication schemes. As opposed to point-to-point communication, PyTorch also supports collective communication across all workers in the group. Typical collective primitives include `Gather`, `Broadcast`, `Reduce`, `AllReduce`, etc.

PyTorch supports three built-in communication backends with different capabilities, i.e., MPI, Gloo,[4] and NCCL.[5] MPI (Message Passing Interface) is a communication standard that is broadly used in areas of high-performance computing and parallel processing. But MPI is not originally compiled in PyTorch's binaries, users need to install PyTorch from the source to include MPI libraries. Gloo is pre-compiled in PyTorch libraries. It supports all the point-to-point and collective communication operators on CPU and collective communication operators on GPU. NCCL is a highly optimized communication backend for collective communication over CUDA tensors. Users who only consider CUDA tensors should always choose NCCL because it achieves the best practical performance and is pre-compiled with PyTorch.

MXNet

MXNet is a scalable machine learning library specially optimized for deep neural networks [37]. It builds deep learning models using declarative symbolic expressions, imperative tensor computation, automatic differentiation, and heterogeneous environment capability.

---

[4] https://github.com/facebookincubator/gloo.

[5] https://github.com/NVIDIA/nccl.

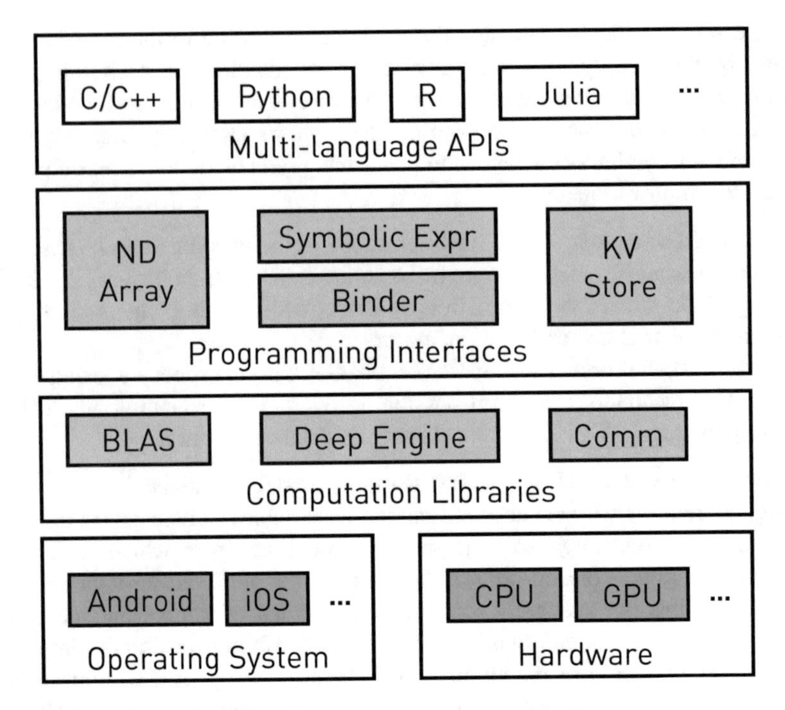

Fig. 4.18 Architecture of MXNet

Programming Interface MXNet supports two programming paradigms—imperative programming and declarative programming. For an expression $a = b+1$, the imperative programming model eagerly computes the result and stores it on a; while the declarative programming model returns a computation graph, binds the data to b and performs the actual computation later. Compared with imperative programming, declarative programming allows the optimization of performance due to deferred execution.

As the system architecture of MXNet in Fig. 4.18 illustrates, the programming interface of MXNet consists of several components, including `Symbol`, `NDArray`, and `KVStore`. We elaborate each component individually below:

- *Symbol: Declarative Symbolic Expressions.* `Symbol`, a symbolic expression, is used to declare the computation graph in MXNet. Symbols are composed of operators, such as mathematical operators (e.g., +) or neural network layers (e.g., convolutional layer). Each operator, which may have internal states, takes one or several inputs and generates outputs. Listing 4.4 shows how to define a neural network symbol by composing the input data and several neural network layers. To execute the *forward* phase of the constructed symbol, one needs to bind the defined variables with data and declare the output. The *backward* phase is internally supported by MXNet using automatic symbolic differentiation. Besides,

Listing 4.4 An Example of Symbol Expression with MXNet.

```
mlp = @mx.chain mx.Variable(:data) =>
  mx.FullyConnected(num_hidden=64) =>
  mx.Activation(act_type=:relu) =>
  mx.FullyConnected(num_hidden=10) =>
  mx.Softmax()
```

MXNet implements additional functions, e.g., load, save, and visualization, for each symbol.

- *NDArray: Imperative Tensor Computation.* The computation of tensors in MXNet is implemented through `NDArray`. NDArray works together with the execution plan declared by `Symbol` and is compatible with the multiple host languages.
- *KVStore: Distributed Data Store.* MXNet is designed to be easily paralleled over distributed devices. Under such a setting, MXNet establishes a distributed key-value store, called KVStore, for data storage and data access across devices. It provides two primitives for users—*push* sends a key-value pair from a device to the KVStore, and *pull* obtains the value of a specific key from the KVStore. KVStore also has synchronization mechanisms, sequential consistency, and eventual consistency, for data needed to be synchronized.

Computation Graph Similar to TensorFlow and PyTorch, the constructed symbolic expression can be represented as a computation graph. In the created computation graph, the input data, the model parameters, and the operators are represented as vertices, and the tensors during execution are represented as edges.

Hierarchical Communication MXNet implements KVStore using the parameter server architecture. MXNet introduces two optimizations w.r.t. the communication across workers. First, the servers are used to schedule the operations on KVStore and handle the data consistency. In this manner, the data synchronization runs independently with the computation. Second, a hierarchical communication architecture is proposed to save communication, as shown in Fig. 4.19. The first level server runs inside a machine and performs synchronization between the devices. The second level servers run in separate machines and perform inter-machine communication. Since the first layer server aggregates the data on colocated devices without being sent to the network, it can save considerable communication overhead. Besides, intra- and inter-machine synchronization can use different consistency models for advanced optimizations.

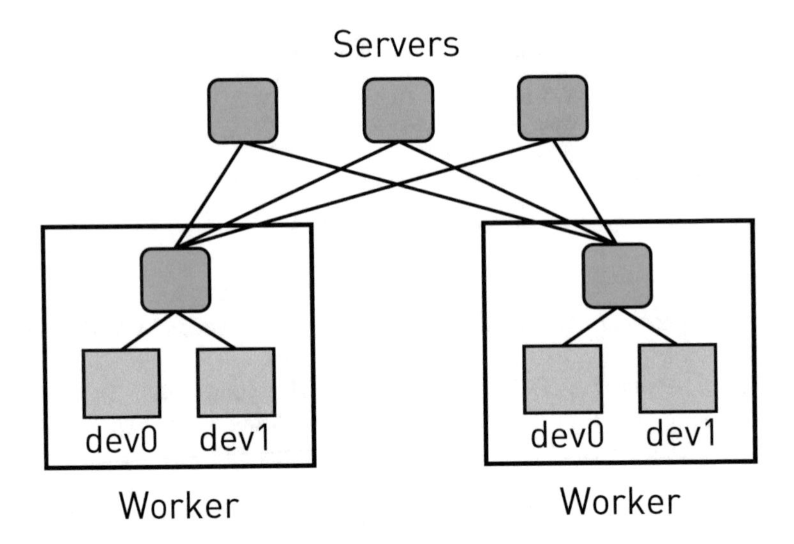

Fig. 4.19 Communication framework of MXNet

4.4 Cloud Machine Learning Systems

In recent years, cloud computing has become increasingly popular due to its flexibility, scalability, pay-by-use, and low maintenance effort. Since machine learning is currently one of the most important applications, there is a trend that deploys machine learning workloads in the cloud [38–40].

Researchers have investigated building machine learning systems over the cloud infrastructure in different ways. Some researchers build machine learning system using rented VMs in the cloud. The problem is how to resolve specific challenges arising in the cloud environments. For example, many international companies deploy all business applications in the cloud, which run in multiple data centers to serve customers all over the world. Since the application data is generated in geo-distributed data centers, training machine learning models over the data is difficult because data communication between data centers is much slower and more expensive than that inside one data center. Some other researchers study how to perform machine learning workloads using novel cloud infrastructures instead of using the traditional rented VMs. Specifically, most cloud providers, such as Google Could, Amazon AWS and Microsoft Azure, offer a new "serverless" computing infrastructure for users. Different from traditional "serverful" computing, which reserves exclusive resources for users and charges for the reserved volume, serverless computing does not reserve any resources and charges for the actual executions. Building distributed machine learning systems over the new infrastructure is nontrivial. This often needs to understand the fundamental characteristics and completely redesign the whole architecture.

In this section, we present several works on these two topics—geo-distributed machine learning system and machine learning system over serverless infrastructure.

4.4.1 Geo-Distributed Systems

Hsieh et al. [41] propose a geo-distributed machine learning system called Gaia. When the available training data are generated all over the world, it is infeasible to transfer the globally distributed data to a centralized data center and train the machine learning model inside the data center using LAN (local area network). The reason is obvious—transmitting large-scale data through WANs (wide area network) can be very slow due to small bandwidth and high latency. Besides, sometimes it is not legal to move data across countries due to privacy-related laws. An alternative that guarantees data privacy is to let each data center calculate the gradients and communicate the gradients across the data centers. However, it still requires expensive communication cost for iteratively transmitting the gradients. Gaia proposes to decouple the communication within a data center from the communication between data centers. Furthermore, Gaia introduces a new synchronization protocol to drop less significant communication between data centers and assure model correctness meanwhile.

System Architecture As shown in Fig. 4.20, Gaia implements a parameter server architecture [16, 18] to perform model synchronization. In the traditional parameter server architecture, several servers together store a global copy of the model parameters, and each worker maintains a local copy. The workers periodically synchronize their local model parameters with each other through the parameter server. However, directly implementing this parameter server architecture in a geo-distributed environment may introduce tremendous communication costs because WAN is generally much slower than LAN. To address this challenge, Gaia proposes a new parameter server architecture that decouples the intra-data-center communication through LANs from the inter-data-center communication through WANs.

In the parameter server architecture of Gaia, each data center launches some machines to act as parameter servers. The parameter servers maintain a copy of the model parameter, called *global model copy*, and the global model copy is partitioned over the servers so that each server stores a model shard. The workers run in the data-parallel mode in which each worker has an independent data shard, each of them holds a local copy of the model parameters. Each worker reads the global model copy from its located data center, performs local computation (e.g., calculate gradient) and updates the global model copy accordingly. Inside one data center, the workers can synchronize with each other using popular protocols, including bulk synchronous parallel (BSP) and stale synchronous parallel (SSP). BSP introduces strict synchronization barriers, while SSP allows aggressive execution on the faster workers.

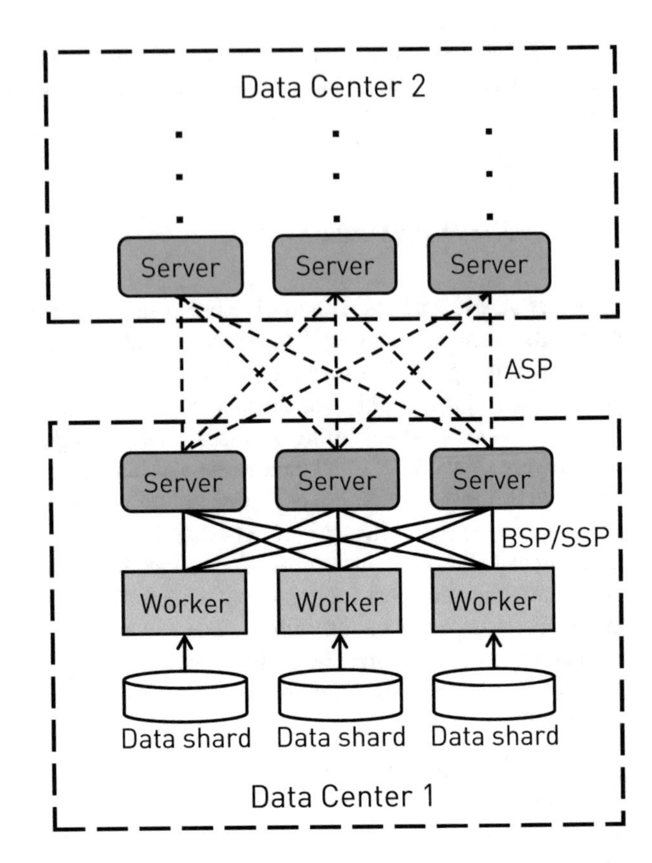

Fig. 4.20 Communication framework of Gaia

To guarantee the correctness of the fitted model, different data centers need to synchronize their global model copies at some point since each individual global model copy may be biased. In terms of synchronization across data centers, Gaia proposes a novel synchronous protocol, called approximate synchronous parallel (ASP), to exchange global model copies on geo-distributed parameter servers. We will elaborate on the details of ASP later.

Approximate Synchronous Protocol Gaia proposes to reduce the communication cost by eliminating "insignificant" communication. The authors state that the updates generated by the workers reveal diverse significances. The majority of updates are less significant, meaning that they barely or slightly improve the model quality.

Based on this insight, Gaia proposes a synchronization protocol, named approximate synchronous parallel (ASP), which lets the parameter servers in each data center communicate only significant updates with other data centers. ASP assures that the global model copy in each data center is approximately correct. There are three key components in ASP: the significance filter, the ASP selective barrier, and the ASP mirror clock.

- *Significance filter.* In the implementation of the significance filter, there is a significance function and an initial significance threshold. The significant function gives a significance value to an update. Then, the update is determined as significant if the significant value is larger than the threshold. A typical significance function uses the magnitude of the update compared with the current model, i.e., $\frac{|update|}{|model|}$. The initial significance threshold can be configured by the users, e.g., 1% and 10%.

  Inside one data center, the parameter server aggregates the updates from local workers. If the update is determined to be significant, the parameter server shares the aggregated updates with other data centers. Otherwise, if the update is determined as insignificant, Gaia lets the parameter server inside one data center accumulate the updates until they become significant as the training proceeds. Furthermore, Gaia decreases the significance threshold across iterations to stabilize the convergence when the current model approaches optimality.
- *ASP selective barrier.* Under a geo-distributed setting, the communication latency between data centers is much larger than that inside each data center due to complex network topology and unpredictable network congestion. As a consequence, some significant updates may arrive late, making different global model copies inconsistency. Gaia proposes a selective barrier to address this problem. If one data center is behind other data centers for receiving significant updates, the parameter server sends ASP selective barrier control messages to other data centers, in which there are the indexes of these significant updates. Other data centers receive the control message and block their local workers from reading the corresponding model parameters until they receive the significant updates from the source of the control message.
- *ASP mirror clock.* Although ASP selective barrier assures that the significant updates are exchanged within a bounded latency. But it assumes that the network latency of WANs is bounded, which is not always true in reality. Gaia engineers a mirror clock in ASP to let the workers know the situation of transferred significant updates in time.

  Each update proposed by a worker is attached with a *clock* (the current running iteration). When the parameter server inside a data center receives all the updates from the local workers, it reports the clock to parameter servers in other data centers. When a parameter server finds that its clock is ahead of the slowest data center by a predefined staleness threshold, the parameter server blocks the reading requests from its local workers until the slowest data center catches up. This synchronization scheme, known as stale synchronous protocol (SSP) [20], achieves synchronization control efficiently, as it only incurs very small messages.

4.4.2 Serverless Systems

When the aforementioned distributed machine learning systems are deployed in the cloud, the most common approach is to use VMs provided by the cloud. In VM-based environments, users have to build a cluster by renting VMs or reserve a cluster with predetermined configuration parameters (e.g., Azure HDInsight [42]). As a result, users pay bills based on the computation resources that have been reserved, regardless of whether these resources are in use or not. Moreover, users have to manage the resources by themselves—there is no elasticity or autoscaling if the reserved computation resources turn out to be insufficient, even for just a short moment (e.g., during the peak of a periodic or seasonal workload). Therefore, to tolerate such uncertainties, users tend to *overprovisioning* by reserving more computation resources than actually needed. In this section, we use the term IaaS (infrastructure as a service) to denote the VM-based infrastructure.

To address the disadvantages of IaaS, a new computing paradigm called serverless computing has recently emerged as a new type of computation infrastructure [43–45]. Serverless computing has been offered by major cloud service providers (e.g., AWS Lambda [46], Azure Functions [47], Google Cloud Functions [48]). Different from IaaS, serverless infrastructure does not reserve resources for users and allocates free resources only when the user program is triggered. In terms of pricing, serverless computing offers a novel "pay by usage" pricing model and can be more cost-effective than traditional IaaS infrastructure that charges users based on the amount of computational resources being reserved. With serverless, the user specifies a *function* that he or she hopes to execute and is charged only for the duration of the function execution. Unlike serverful computing, i.e., IaaS, which is inefficient in scalability, the users can also easily scale up the computation by specifying the number of such functions that are executed concurrently. In summary, the move toward FaaS infrastructures lifts the burden of managing computational resources from users, improves the resource utilization due to on-demand resource allocation and offers more flexible charging in case the program execution is infrequent. In this section, we also use the term FaaS (function as a service) to denote the serverless infrastructure.

Serverless computing was initially developed for web microservices and IoT scenarios, and has been favored by various applications, such as event processing, API composition, API aggregation, and data-flow control [44, 49]. It can effectively lift the burden of provisioning and managing cloud computation resources (e.g., with autoscaling) from application developers. Recently, some researchers have explored the capability of serverless computing in data-intensive applications, which stimulates intensive interests in the data management community [50–54]. Previous work has shown that adopting a serverless infrastructure for certain types of workloads, ranging from ETL [55] to analytical queries over cold data [56, 57], can significantly lower the cost. These data management workloads benefit from FaaS by taking advantage of the unlimited elasticity, pay per use, and lower start-up overhead provided by a serverless infrastructure.

As the research goes further, some works try to study the potential of FaaS on machine learning workloads as machine learning applications grow. FaaS is naturally suitable for machine learning inference [58] because an inference workload can be deployed on multiple workers, and each worker runs independently. However, when serverless computing is used to train machine learning models, the current offerings of major cloud service providers (e.g., AWS Lambda, Azure Functions, Google Cloud Functions) impose certain limitations and/or constraints that shed some of the values by shifting from IaaS to FaaS infrastructures. First, the current FaaS infrastructures only support *stateless* function calls with limited computation resource and duration. For instance, a function call in AWS Lambda can use up to 3 GB of memory and must finish within 15 min [59]. Second, the running instances in FaaS infrastructure are stateless, so that the current infrastructure does not support point-to-point communications between the running instances. Such constraints prohibit the implementation of FaaS-based machine learning systems by directly migrating some simple yet natural ideas from IaaS-based machine learning systems. For example, one cannot just wrap the code of SGD in an AWS Lambda function and execute it, which would easily run out of memory or hit the timeout limit on large training data. Indeed, the imperfection of state-of-the-art FaaS offerings raises many new challenges for designing machine learning systems and leads to a rich design space. Below, we present several reported works on developing machine learning systems over the FaaS (serverless) infrastructure [60].

Cirrus

Carreira et al. [61] design Cirrus, an end-to-end machine learning system over Lambda, the serverless infrastructure of Amazon AWS. Cirrus has four major building blocks—a Python frontend, a client-side backend, a serverless runtime, and a distributed data store. We describe the principle and detail of each building block individual.

- *Python frontend.* The frontend of Cirrus is implemented with Python, the most popular programming language for data scientists. The frontend hides the low-level implementations so that the users do not need to deal with the underlying system details. The Python API contains three submodules—preprocessing, training, and hyperparameter optimization. The preprocessing submodule provides transformations for the input dataset stored in Amazon S3, such as feature scaling, standardization, and hashing. The training submodule provides popular machine learning models trained with stochastic gradient descent, including Logistic Regression, Latent Dirichlet Allocation, and Matrix Factorization. The hyperparameter optimization submodule uses a grid search to find the optimal hyperparameters (e.g., learning rate, regularization, batch size) over a set of candidates.
- *Client-side backend.* The backend at the client side is responsible for some tasks between the Python API and the serverless backend, e.g., parsing the training dataset, loading the dataset to S3, launching the Cirrus workers in Lambda, and maintaining the computation progress.

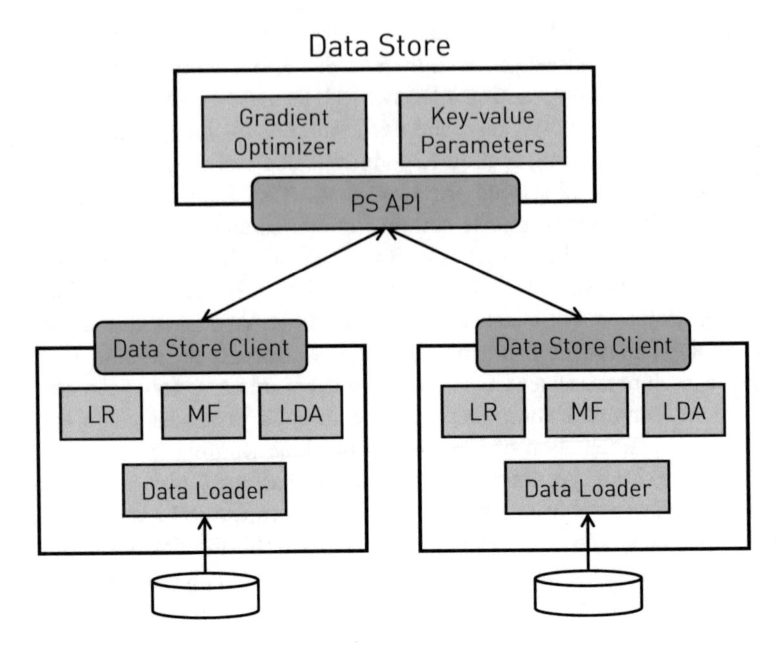

Fig. 4.21 Architecture of cirrus

- *Serverless runtime.* The actual runtime is implemented in the serverless infrastructure, consisting of basic primitives of computation and communication. Specifically, the serverless runtime provides two interfaces. The first interface is an iterator that is provided for manipulating the training dataset from S3. It can prefetch and store minibatch data in a buffer to alleviate the high latency of S3 (typically > 10 ms). The second interface is a handler for accessing data in the distributed data store. It provides data compression, sparse data formatting, asynchronous communication, and data sharding.
- *Distribute data store.* Recalling the disadvantage of serverless infrastructure, the running workers cannot communicate with each other due to their stateless nature. As shown in Fig. 4.21, a distributed data store is established to store intermediate data shared across all the running workers. Cirrus uses several VMs to implement the data store to achieve low communication latency.

SIREN

Wang et al. [62] develop an asynchronous distributed machine learning system SIREN based on the serverless infrastructure. Figure 4.22 shows the architecture of SIREN. The code package of one user is uploaded to the cloud platform, typically through a web UI. The cloud provider also offers many popular libraries for users, which can be configured on the UI. When the user code is triggered by the user or some upstream notification, the serverless cloud platform (Lambda in this work) launches a set of stateless functions according to the initial configurations, e.g.,

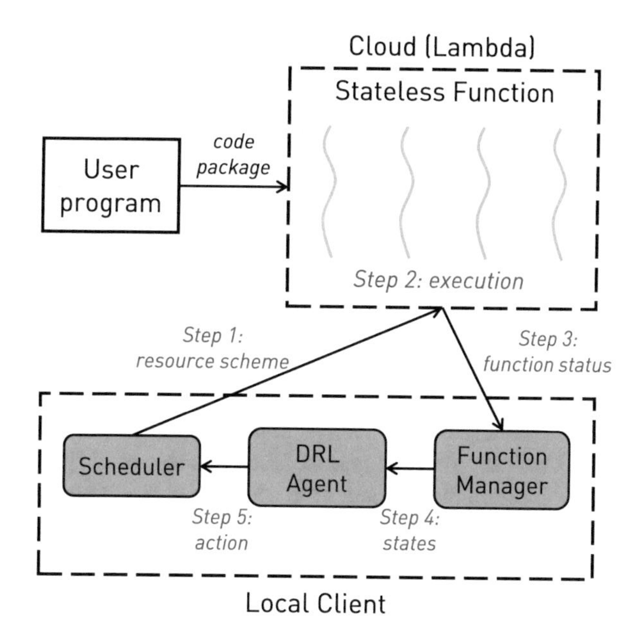

Fig. 4.22 Architecture of SIREN

the number of concurrent functions and the memory size. SIREN adopts the data-parallel training scheme in which the stateless functions process different batches of data. Once an iteration is finished, the function status and statistics are sent to a local client. The local client has a local client that schedules the serverless computing with a deep reinforcement learning (DRL) agent. The DRL agent receives function status from the stateless functions, updates the decision model, and calculates the action taken for the next iteration. Based on the actions made by the DRL agent, the scheduler adaptively adjusts the resource used for the stateless functions, including the number of functions and the memory configurations.

SIREN implements a parameter server to aggregate the model parameters from all the stateless functions. Nevertheless, the parameter server architecture in SIREN is different from the traditional parameter servers that run in VMs. SIREN chooses to store the data and model in a data store. e.g., Amazon S3, to overcome the lack of direct communication between stateless functions. A global copy of the model parameters is stored in the data store, which is shared by all the functions. At each iteration, each function reads the current model parameters from the data store, computes gradients using its local training data and directly updates the model in the data store using the generated gradients. The consistency model of SIREN uses a hybrid synchronous parallel (HSP). Within one epoch, all the functions run asynchronously and update the global model parameters. To guarantee the correct convergence, HSP sets a synchronization barrier at the end of each epoch, where the DRL agent collects the function statistics and makes the scheduling decision for the next epoch.

Other Serverless Systems

Several other works also study how to build machine learning systems in the serverless environments. We briefly summarize these works here since they use similar techniques mentioned in the above two systems. Feng et al. [63] propose a distributed system for training neural networks in serverless environments. It adopts data parallelism that partitions the training dataset across serverless workers. The global model parameters and the gradients generated by the workers are stored in a data store such as Amazon S3. One worker is chosen to act as a parameter server that aggregates the gradients on the data store and updates the global model parameters. Gupta et al. [64] propose training machine learning models in serverless infrastructure using a second-order gradient optimization algorithm. Numpywren [65] is a linear algebra library build upon the serverless framework, supporting algorithms such as matrix multiplication, singular value decomposition, and Cholesky decomposition. Bhattacharjee et al. [66] design a serverless framework Stratum that manages the lifecycle of machine learning workloads, consisting of data ingestion tools, streaming applications, analytical tools, inference jobs, and visualization tools.

4.5 In-Database Machine Learning Systems

The aforementioned machine learning systems assume that the training datasets are stored in distributed file systems (e.g., HDFS and Amazon S3). But this is not the case for all the workloads. Many real-world applications generate relational data and choose to store them in databases. According to a recent survey of Kaggle, relational data are the most common type of data for data scientists.[6] If the traditional machine learning systems are chosen to train machine learning models over the training dataset stored in a database, one needs to first select relevant data entries from the database using SQL query and move them to an external data store. This may incur extensive costs on data movement across system boundaries. Worse, sometimes moving data out of the database is infeasible due to privacy protection regulations. An intuitive solution to address this issue is to train machine learning models inside the database without moving data out, which stimulates the development of in-database machine learning systems [67–69].

MADlib

MADlib is an open-source in-database data analytic system that provides machine learning functionalities based on SQL operators [70]. It implements a wide variety of model learning models, including linear regression, logistic regression, naive Bayes, decision tree, support vector machine, KMeans, LDA, and many statistical models.

[6] https://www.kaggle.com/surveys/2017.

Programming Interfaces The optimization of machine learning models can be fundamentally expressed as a series of mathematical linear algebra operations over vectors or matrices, such as matrix multiplication and dot products. To implement linear algebra operations over relational databases, the design space of MADlib has several major considerations. First, the data matrices in a relational database must be properly partitioned into chunks, ensuring that a single node has enough memory to hold one chunk. Each chunk is attached with a key, enabling SQL operators to access the chunks across the machines. Secondly, given the partitioned matrices, MADlib tries to perform efficient linear algebra computations over the chunks.

MADlib has offered programming primitives, including data management and linear algebra computation, for implementing machine learning models in databases.

- *User-defined aggregation.* A basic programming interface is user-defined aggregation (UDAs). It is used to implement mathematical operators that take as input the number of rows and run in the data-parallel scheme. Typically, a user-defined aggregation function consists of several user-defined functions.

  - *Transition function.* The current transition state and a new data point are fed to the transition function. They are combined into a new transition state.
  - *Merge function.* Two transition states are input into a merge function to generate a new transition state.
  - *Final function.* A transition state is transformed into the output using a final function.

  Data parallel can be easily implemented using the above programming abstractions if the transition function is associative and the merge function returns consistent outputs for repetitive executions. However, user-defined aggregation alone is not enough for implementing machine learning models. First, the optimization algorithms used to train machine learning models are usually iterative. They need to pass the training dataset many times until convergence. At the end of each iteration, a coordinator is required to synchronize the states of data-parallel workers. Second, the logic of an SQL query is generally fixed, that is, the schema of the input tables and the output tables must be fixed beforehand. These limitations are resolved by user-defined driver functions and templated queries described below.
- *Driver function.* The iterative processing of machine learning models is achieved by the driver functions. To drive the independent states in n iterations, MADlib simply declares a virtual tale with n rows and joins it with a view representing a single iteration. If different iterations are correlated, e.g., the current iteration depends on the previous function, MADlib utilizes the window aggregation function to share states across iterations [71]. As for recursive executions (e.g., for statement or loop), MADlib uses the recursion feature of SQL to handle iterative algorithms with stopping conditions. To ease the programming, MADlib implements iterative algorithms by writing a driver UDF (user-defined function) in Python. The driver code writes inter-iteration outputs into a temporary table that is reused in the subsequent iterations.

- *Templated queries.* The problem of the fixed schema of SQL queries is addressed by writing "templated" queries that can work for arbitrary schemas and enable the later filling of arity, column names, and data types.
- *Data representations and inner loops.* In addition to the operations on chunks or multiple rows, MADlib also implements efficient row-level operations. The row-level UDFs are implemented with C or C++, which calls open-source libraries such as LAPACK and Eigen when computing dense matrix operations. For sparse matrix manipulations, since the existing libraries cannot handle them efficiently, MADlib implements a sparse matrix library in C with a run-length encoding scheme.
- *C++ abstraction layer.* MADlib provides a C++ abstraction layer to ease the writing of high-performance UDFs and to embed DBMS-specific logic inside the abstraction layer. This C++ abstraction provides three classes of functionality—(1) type bridging between C++ primitives (e.g., data type and method) and database primitives, (2) resource management shims between C++ runtime and DBMS-managed memory interfaces, and (3) math library integration which makes the programming easier for the users.

Bismarck

Feng et al. propose BISMARCK, a unified architecture for in-database analytics [72]. With this unified architecture, analytic models can be trained inside the database, and users do not need to implement new statistical techniques from scratch which require new memory requirements, new data accessing methods, etc. Bismarck focuses on machine learning models trained by incremental gradient descent (IGD), which approximates the full gradient using a single instance. Particularly, Bismarck designs techniques to solve two key factors affecting the performance—the order of the data and the parallelization on a single multicore machine.

Architecture of Bismarck The brief architecture of Bismarck is shown in Fig. 4.23. One user calls a machine learning model in the library of Bismarck and sets necessary specifications (e.g., input data and hyperparameters). The Bismarck engine receives the specifications from the user and runs the task using incremental gradient descent (IGD). To implement IGD inside a DBMS with data accessing pattern similar to SQL queries, Bismarck uses user-defined aggregate (UDA) supported by most DBMS. The UDA mechanism is leveraged to perform gradient computation and loss computation.

The UDA for IGD contains three key functions—`initialize(state)`, `transition(state, data)`, and `terminate(state)`. The specific names of these functions may vary for different databases, for example, PostgreSQL names them "initcond," "sfunc," and "finalfunc" [73]. The `state` variable denotes the model parameters in Bismarck and some metadata (e.g., number of iterations), and the `data` variable denotes a training tuple in the table.

- The initialization function initializes the model parameters with the given values or an existing model returned by a previous execution.

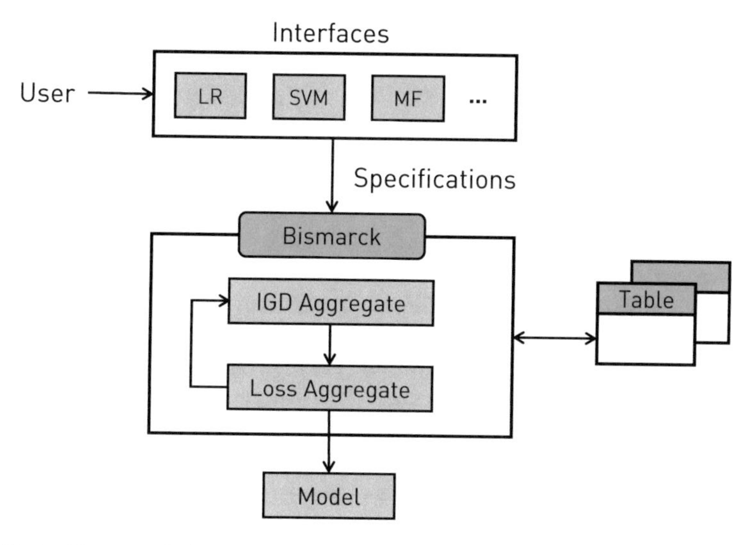

Fig. 4.23 Architecture of Bismarck

- The transition function computes the gradient over the input `data` instance, and updates the current model parameters. This function may be different for various machine learning models since each individual model has a unique objective function and pattern of gradient computation. Users only need to write a few lines of customized code in this function to design a new analytical model, while reusing the rest of the architecture across diverse models.
- The termination functions stop the gradient computation and return the model.

Since the order of training data often influences the output model, Bismarck performs reordering or sampling for the training data to improve the convergence performance of IGD. This can be achieved by operators in popular DBMSs, e.g., `ORDER_BY_RANDOM` operator in PostgreSQL.

To support the iterative execution of gradient optimization algorithms, Bismarck requires to know the number of iterations. Bismarck calls a Boolean function to support this functionality. The iterative criteria can be running for a fixed number of iterations, the norm of computed gradients, or the value of the statistical loss.

To calculate the value of the objective function (a.k.a. *loss*), typically performed at the end of each iteration or epoch, Bismarck implements the computation using a UDA similar to the scheme of IGD UDA.

Data Reordering A special optimization of Bismarck is to perform a reordering for the training data. Empirically, certain data ordering may make the machine learning model converge in fewer iterations. Although traditional machine learning workloads do not reorder the training data since distributed data shards are generally considered i.i.d (independent and identically distributed), this is not the case for many datasets stored in databases. The authors state that data inside a database is often clustered for purposes unrelated to the analytical task. For instance, classification datasets may be clustered by the class labels. To improve the convergence performance for in-database datasets, Bismarck randomly shuffles the data once before the training. In this manner, a potential pathological ordering in data stored in a database is avoided. Bismarck does not choose to shuffle the data repetitively (e.g., at each epoch); therefore, the shuffling overhead does not significantly affect the overall performance.

Parallelization of Gradient Computation The major computation in training a machine learning model is the gradient computation, i.e., IGD aggregate computation in Bismarck. Bismarck proposes parallelizing IGD aggregates to speed up the performance through two mechanisms—(1) using standard UDA features and (2) using shared-memory features. The pure UDA solution benefits from the "shared-nothing" parallelism in UDAs provided by most DBMSes. The shared-memory UDA is also provided by most DBMSs in which the global model parameters are stored in shared-memory. The parallel threads run on different data shards, read and update the model parameters.

Other In-Database Machine Learning Systems
Other works develop in-database machine learning engines using similar techniques proposed in the above systems for specific scenarios or models. Khamis et al. [74] propose an in-database machine learning system, specially optimized for sparse scenarios. The system can handle both continuous features and categorical features which are represented as one-hot encoding. Through defining the representations of sparse tensors and the gradient computation over them, the system achieves lower space and time complexities. Sandha and Cabrera demonstrate a distributed machine learning system natively inside database engines [75]. The system is built with the Teradata SQL engine [76], which runs in parallel over shared-nothing architecture. Schleich et al. [67] target training linear regression models over datasets obtained by join queries on database tables. Chen et al. [68] also study datasets generated by joining tables. To present manual rewriting of machine learning techniques, the authors propose to use a formal linear algebra to represent the computations in machine learning. They also devise a data representation for normalized data and rewrite rules that transform operations on denormalized data into operations on normalized data. DB4ML [77] is an in-memory database kernel that allows the implementation of user-defined machine learning models. A novel programming model is designed to support iterative transactions that can be extended for various machine learning models.

References

1. Dean, Jeffrey and Ghemawat, Sanjay: MapReduce: simplified data processing on large clusters. Communications of the ACM. 51(1), 107–113 (2008)
2. Apache Hadoop, https://hadoop.apache.org/

3. Zaharia, Matei and Chowdhury, Mosharaf and Das, Tathagata and Dave, Ankur and Ma, Justin and McCauly, Murphy and Franklin, Michael J and Shenker, Scott and Stoica, Ion: Resilient distributed datasets: A fault-tolerant abstraction for in-memory cluster computing. Presented as part of the 9th USENIX Symposium on Networked Systems Design and Implementation (NSDI 12). 15–28 (2012)
4. Zaharia, Matei and Xin, Reynold S and Wendell, Patrick and Das, Tathagata and Armbrust, Michael and Dave, Ankur and Meng, Xiangrui and Rosen, Josh and Venkataraman, Shivaram and Franklin, Michael J and others: Apache spark: a unified engine for big data processing. Communications of the ACM. 59(11), 56–65 (2016)
5. Apache Mahout, http://mahout.apache.org/
6. Meng, Xiangrui and Bradley, Joseph and Yavuz, Burak and Sparks, Evan and Venkataraman, Shivaram and Liu, Davies and Freeman, Jeremy and Tsai, DB and Amde, Manish and Owen, Sean and others: Mllib: Machine learning in Apache Spark. The Journal of Machine Learning Research. 17(1), 1235–1241 (2016)
7. Bottou, Léon: Large-scale machine learning with stochastic gradient descent. Proceedings of COMPSTAT'2010. 177–186 (2010)
8. Broyden, Charles George: The convergence of a class of double-rank minimization algorithms 1. general considerations. IMA Journal of Applied Mathematics. 6(1), 76–90 (1970)
9. Liu, Dong C and Nocedal, Jorge: On the limited memory BFGS method for large scale optimization. Mathematical Programming. 45(1-3), 503–528 (1989)
10. Nocedal, Jorge: Updating quasi-Newton matrices with limited storage. Mathematics of Computation. 35(151), 773–782 (1980)
11. Andrew, Galen and Gao, Jianfeng: Scalable training of L 1-regularized log-linear models. Proceedings of the 24th International Conference on Machine Learning. 33–40 (2007)
12. Panda, Biswanath and Herbach, Joshua S and Basu, Sugato and Bayardo, Roberto J: Planet: massively parallel learning of tree ensembles with MapReduce. (2009)
13. Smola, Alexander and Narayanamurthy, Shravan: An architecture for parallel topic models. Proceedings of the VLDB Endowment. 3(1-2), 703–710 (2010)
14. Blei, David M and Ng, Andrew Y and Jordan, Michael I: Latent Dirichlet allocation. Journal of machine Learning research. 3, 993–1022 (2003)
15. Smyth, Padhraic and Welling, Max and Asuncion, Arthur: Asynchronous distributed learning of topic models. Advances in Neural Information Processing Systems. 21, 81–88 (2008)
16. Dean, Jeffrey and Corrado, Greg and Monga, Rajat and Chen, Kai and Devin, Matthieu and Mao, Mark and Ranzato, Marc'aurelio and Senior, Andrew and Tucker, Paul and Yang, Ke and others: Large scale distributed deep networks. Advances in Neural Information Processing Systems. 1223–1231 (2012)
17. Duchi, John and Hazan, Elad and Singer, Yoram: Adaptive subgradient methods for online learning and stochastic optimization. Journal of Machine Learning Research. 12(7) (2011)
18. Xing, Eric P and Ho, Qirong and Dai, Wei and Kim, Jin Kyu and Wei, Jinliang and Lee, Seunghak and Zheng, Xun and Xie, Pengtao and Kumar, Abhimanu and Yu, Yaoliang: Petuum: A new platform for distributed machine learning on big data. IEEE Transactions on Big Data. 1(2), 49–67 (2015)
19. Zinkevich, Martin and Weimer, Markus and Li, Lihong and Smola, Alex J: Parallelized stochastic gradient descent. Advances in Neural Information Processing Systems. 2595–2603 (2010)
20. Ho, Qirong and Cipar, James and Cui, Henggang and Lee, Seunghak and Kim, Jin Kyu and Gibbons, Phillip B and Gibson, Garth A and Ganger, Greg and Xing, Eric P: More effective distributed ml via a stale synchronous parallel parameter server. Advances in Neural Information Processing Systems. 1223–1231 (2013)
21. Dai, Wei and Kumar, Abhimanu and Wei, Jinliang and Ho, Qirong and Gibson, Garth and Xing, Eric P: High-performance distributed ML at scale through parameter server consistency models. arXiv preprint arXiv:1410.8043. (2014)
22. Li, Mu and Andersen, David G and Park, Jun Woo and Smola, Alexander J and Ahmed, Amr and Josifovski, Vanja and Long, James and Shekita, Eugene J and Su, Bor-Yiing:

Scaling distributed machine learning with the parameter server. 11th USENIX Symposium on Operating Systems Design and Implementation (OSDI 14). 583–598 (2014)
23. Karger, David and Sherman, Alex and Berkheimer, Andy and Bogstad, Bill and Dhanidina, Rizwan and Iwamoto, Ken and Kim, Brian and Matkins, Luke and Yerushalmi, Yoav: Web caching with consistent hashing. Computer Networks. 31(11-16), 1203–1213 (1999)
24. Byers, John and Considine, Jeffrey and Mitzenmacher, Michael: Simple load balancing for distributed hash tables. International Workshop on Peer-to-Peer Systems. 80–87 (2003)
25. Jiang, Jie and Yu, Lele and Jiang, Jiawei and Liu, Yuhong and Cui, Bin: Angel: a new large-scale machine learning system. National Science Review. 5(2), 216–236 (2018)
26. https://github.com/Angel-ML/angel
27. Paszke, Adam and Gross, Sam and Massa, Francisco and Lerer, Adam and Bradbury, James and Chanan, Gregory and Killeen, Trevor and Lin, Zeming and Gimelshein, Natalia and Antiga, Luca and others: Pytorch: An imperative style, high-performance deep learning library. Advances in Neural Information Processing Systems. 8026–8037 (2019)
28. Jiang, Jiawei and Xiao, Pin and Yu, Lele and Li, Xiaosen and Cheng, Jiefeng and Miao, Xupeng and Zhang, Zhipeng and Cui, Bin: PSGraph: How Tencent trains extremely large-scale graphs with Spark? 2020 IEEE 36th International Conference on Data Engineering (ICDE) 1549–1557, (2020)
29. Chen, Tianqi and Guestrin, Carlos: Xgboost: A scalable tree boosting system. Proceedings of the 22nd ACM SIGKDD International Conference on Knowledge Discovery and Data Mining. 785–794 (2016)
30. Greenwald, Michael and Khanna, Sanjeev: Space-efficient online computation of quantile summaries. ACM SIGMOD Record. 30(2), 58–66 (2001)
31. Zhang, Qi and Wang, Wei: A fast algorithm for approximate quantiles in high speed data streams. 19th International Conference on Scientific and Statistical Database Management (SSDBM 2007). 29–29 (2007)
32. Ke, Guolin and Meng, Qi and Finley, Thomas and Wang, Taifeng and Chen, Wei and Ma, Weidong and Ye, Qiwei and Liu, Tie-Yan: Lightgbm: A highly efficient gradient boosting decision tree. Advances in Neural Information Processing Systems. 3146–3154 (2017)
33. Meng, Qi and Ke, Guolin and Wang, Taifeng and Chen, Wei and Ye, Qiwei and Ma, Zhi-Ming and Liu, Tieyan: A communication-efficient parallel algorithm for decision tree. Advances in Neural Information Processing Systems. 1279–1287 (2016)
34. Jia, Yangqing and Shelhamer, Evan and Donahue, Jeff and Karayev, Sergey and Long, Jonathan and Girshick, Ross and Guadarrama, Sergio and Darrell, Trevor: Caffe: Convolutional architecture for fast feature embedding. Proceedings of the 22nd ACM international conference on Multimedia. 675–678 (2014)
35. Abadi, Martín and Barham, Paul and Chen, Jianmin and Chen, Zhifeng and Davis, Andy and Dean, Jeffrey and Devin, Matthieu and Ghemawat, Sanjay and Irving, Geoffrey and Isard, Michael and others: TensorFlow: A system for large-scale machine learning. 12th USENIX symposium on operating systems design and implementation (OSDI 16). 265–283 (2016)
36. Chen, Jianmin and Pan, Xinghao and Monga, Rajat and Bengio, Samy and Jozefowicz, Rafal: Revisiting distributed synchronous SGD. arXiv preprint arXiv:1604.00981. (2016)
37. Chen, Tianqi and Li, Mu and Li, Yutian and Lin, Min and Wang, Naiyan and Wang, Minjie and Xiao, Tianjun and Xu, Bing and Zhang, Chiyuan and Zhang, Zheng: Mxnet: A flexible and efficient machine learning library for heterogeneous distributed systems. arXiv preprint arXiv:1512.01274. (2015)
38. AI Platform in Google Cloud, https://cloud.google.com/ai-platform
39. Azure Machine Learning, https://azure.microsoft.com/en-us/services/machine-learning/
40. Amazon SageMaker, https://aws.amazon.com/cn/sagemaker/
41. Hsieh, Kevin and Harlap, Aaron and Vijaykumar, Nandita and Konomis, Dimitris and Ganger, Gregory R and Gibbons, Phillip B and Mutlu, Onur: Gaia: Geo-distributed machine learning approaching LAN speeds. 14th USENIX Symposium on Networked Systems Design and Implementation (NSDI 17). 629–647 (2017)
42. Azure HDInsight, https://docs.microsoft.com/en-us/azure/hdinsight/

43. Jonas, Eric and Schleier-Smith, Johann and Sreekanti, Vikram and Tsai, Chia-Che and Khandelwal, Anurag and Pu, Qifan and Shankar, Vaishaal and Carreira, Joao and Krauth, Karl and Yadwadkar, Neeraja and others: Cloud programming simplified: A Berkeley view on serverless computing. arXiv preprint arXiv:1902.03383. (2019)
44. Baldini, Ioana and Castro, Paul and Chang, Kerry and Cheng, Perry and Fink, Stephen and Ishakian, Vatche and Mitchell, Nick and Muthusamy, Vinod and Rabbah, Rodric and Slominski, Aleksander and others: Serverless computing: Current trends and open problems. Research Advances in Cloud Computing. 1–20 (2017)
45. McGrath, Garrett and Brenner, Paul R: Serverless computing: Design, implementation, and performance. 2017 IEEE 37th International Conference on Distributed Computing Systems Workshops (ICDCSW). 405–410 (2017)
46. AWS Lambda, https://aws.amazon.com/lambda/
47. Azure Functionsa, https://azure.microsoft.com/en-us/services/functions/
48. Google Cloud Functions, https://cloud.google.com/functions/
49. Hendrickson, Scott and Sturdevant, Stephen and Harter, Tyler and Venkataramani, Venkateshwaran and Arpaci-Dusseau, Andrea C and Arpaci-Dusseau, Remzi H: Serverless computation with OpenLambda. 8th USENIX Workshop on Hot Topics in Cloud Computing (HotCloud 16). (2016)
50. Klimovic, Ana and Wang, Yawen and Stuedi, Patrick and Trivedi, Animesh and Pfefferle, Jonas and Kozyrakis, Christos: Pocket: Elastic ephemeral storage for serverless analytics. 13th USENIX Symposium on Operating Systems Design and Implementation (OSDI 18). 427–444 (2018)
51. Pu, Qifan and Venkataraman, Shivaram and Stoica, Ion: Shuffling, fast and slow: Scalable analytics on serverless infrastructure. 16th USENIX Symposium on Networked Systems Design and Implementation (NSDI 19). 193–206 (2019)
52. Wang, Liang and Li, Mengyuan and Zhang, Yinqian and Ristenpart, Thomas and Swift, Michael: Peeking behind the curtains of serverless platforms. 2018 USENIX Annual Technical Conference (USENIX ATC 18). 133–146 (2018)
53. Akkus, Istemi Ekin and Chen, Ruichuan and Rimac, Ivica and Stein, Manuel and Satzke, Klaus and Beck, Andre and Aditya, Paarijaat and Hilt, Volker: SAND: Towards High-Performance Serverless Computing. 2018 USENIX Annual Technical Conference (USENIX ATC 18). 923–935 (2018)
54. Rausch, Thomas and Hummer, Waldemar and Muthusamy, Vinod and Rashed, Alexander and Dustdar, Schahram: Towards a serverless platform for edge AI. 2nd USENIX Workshop on Hot Topics in Edge Computing (HotEdge 19). (2019)
55. Fingler, Henrique and Akshintala, Amogh and Rossbach, Christopher J: USETL: Unikernels for serverless extract transform and load why should you settle for less? Proceedings of the 10th ACM SIGOPS Asia-Pacific Workshop on Systems. 23–30 (2019)
56. Müller, Ingo and Marroquín, Renato and Alonso, Gustavo: Lambada: Interactive Data Analytics on Cold Data Using Serverless Cloud Infrastructure. Proceedings of the 2020 ACM SIGMOD International Conference on Management of Data. 115–130 (2020)
57. Perron, Matthew and Castro Fernandez, Raul and DeWitt, David and Madden, Samuel: Starling: A Scalable Query Engine on Cloud Functions. Proceedings of the 2020 ACM SIGMOD International Conference on Management of Data. 131–141 (2020)
58. Ishakian, Vatche and Muthusamy, Vinod and Slominski, Aleksander: Serving deep learning models in a serverless platform. 2018 IEEE International Conference on Cloud Engineering (IC2E). 257–262 (2018)
59. AWS Lambda Limitations, https://docs.aws.amazon.com/lambda/latest/dg/gettingstarted-limits.html
60. Carreira, Joao and Fonseca, Pedro and Tumanov, Alexey and Zhang, Andrew and Katz, Randy: A case for serverless machine learning. Workshop on Systems for ML and Open Source Software at NeurIPS. (2018)

61. Carreira, Joao and Fonseca, Pedro and Tumanov, Alexey and Zhang, Andrew and Katz, Randy: Cirrus: a Serverless Framework for End-to-end ML Workflows. Proceedings of the ACM Symposium on Cloud Computing. 13–24 (2019)
62. Wang, Hao and Niu, Di and Li, Baochun: Distributed machine learning with a serverless architecture. IEEE INFOCOM. 1288–1296 (2019)
63. Feng, Lang and Kudva, Prabhakar and Da Silva, Dilma and Hu, Jiang: Exploring serverless computing for neural network training. 2018 IEEE 11th International Conference on Cloud Computing (CLOUD). 334–341 (2018)
64. Gupta, Vipul and Kadhe, Swanand and Courtade, Thomas and Mahoney, Michael W and Ramchandran, Kannan: Oversketched newton: Fast convex optimization for serverless systems. arXiv preprint arXiv:1903.08857. (2019)
65. Shankar, Vaishaal and Krauth, Karl and Pu, Qifan and Jonas, Eric and Venkataraman, Shivaram and Stoica, Ion and Recht, Benjamin and Ragan-Kelley, Jonathan: Numpywren: Serverless linear algebra. arXiv preprint arXiv:1810.09679. (2018)
66. Bhattacharjee, Anirban and Barve, Yogesh and Khare, Shweta and Bao, Shunxing and Gokhale, Aniruddha and Damiano, Thomas: Stratum: A serverless framework for the lifecycle management of machine learning-based data analytics tasks. 2019 USENIX Conference on Operational Machine Learning (OpML 19). 59–61 (2019)
67. Schleich, Maximilian and Olteanu, Dan and Ciucanu, Radu: Learning linear regression models over factorized joins. Proceedings of the 2016 International Conference on Management of Data. 3–18 (2016)
68. Chen, Lingjiao and Kumar, Arun and Naughton, Jeffrey and Patel, Jignesh M: Towards linear algebra over normalized data. arXiv preprint arXiv:1612.07448. (2016)
69. Färber, Franz and Cha, Sang Kyun and Primsch, Jürgen and Bornhövd, Christof and Sigg, Stefan and Lehner, Wolfgang: SAP HANA database: data management for modern business applications. 40(4), 45–51 (2012)
70. Hellerstein, Joe and Ré, Christopher and Schoppmann, Florian and Wang, Daisy Zhe and Fratkin, Eugene and Gorajek, Aleksander and Ng, Kee Siong and Welton, Caleb and Feng, Xixuan and Li, Kun and others: The MADlib analytics library or MAD skills, the SQL. arXiv preprint arXiv:1208.4165. (2012)
71. Wang, Daisy Zhe and Franklin, Michael J and Garofalakis, Minos and Hellerstein, Joseph M and Wick, Michael L: Hybrid in-database inference for declarative information extraction. Proceedings of the 2011 ACM SIGMOD International Conference on Management of data. 517–528 (2011)
72. Feng, Xixuan and Kumar, Arun and Recht, Benjamin and Ré, Christopher: Towards a unified architecture for in-RDBMS analytics. Proceedings of the 2012 ACM SIGMOD International Conference on Management of Data. 325–336 (2012)
73. PostgreSQL, https://www.postgresql.org/
74. Abo Khamis, Mahmoud and Ngo, Hung Q and Nguyen, XuanLong and Olteanu, Dan and Schleich, Maximilian: In-database learning with sparse tensors. Proceedings of the VLDB Endowment. 325–340 (2018)
75. Sandha, Sandeep Singh and Cabrera, Wellington and Al-Kateb, Mohammed and Nair, Sanjay and Srivastava, Mani: In-database distributed machine learning: demonstration using Teradata SQL engine. Proceedings of the VLDB Endowment. 12(12), 1854–1857 (2019)
76. Teradata SQL, https://docs.teradata.com/
77. Jasny, Matthias and Ziegler, Tobias and Kraska, Tim and Roehm, Uwe and Binnig, Carsten: DB4ML-An In-Memory Database Kernel with Machine Learning Support. Proceedings of the 2020 ACM SIGMOD International Conference on Management of Data. 159–173 (2020)

Chapter 5
Conclusion

5.1 Summary of the Book

Machine learning has been the driving force of many real-world applications. Many machine learning models rely on gradient optimization algorithms that fit the model parameters over the training data. As the data volume becomes larger and larger, extending gradient optimization algorithms to distributed environments is indispensable. This book thereby studies gradient optimization in the setting of distributed machine learning.

Based on the anatomy of the routine of distributed machine learning, we first introduce several fundamental building blocks, including parallelism strategies, parameter sharing architecture, synchronization protocol, and communication optimizations. Afterward, we give an overview of the state-of-the-art distributed gradient optimization algorithms, categorized by the formalizations of machine learning models—linear models, neural network models, and tree models. Lastly, we present popular distributed machine learning systems, ranging from general systems, specialized systems, deep learning systems, and systems designed in the cloud and databases.

5.2 Further Reading

There are several other topics related to distributed machine learning. Since they are not the key techniques in designing distributed machine learning methods, we do not elaborate on these topics in this book. Instead, we point out these topics, describe their concepts, and provide further readings for interested readers.

Federated Learning Traditional machine learning methods normally assume the dataset is stored in data centers. However, as the rapid growth of edge devices, such

J. Jiang et al., *Distributed Machine Learning and Gradient Optimization*, Big Data Management, https://doi.org/10.1007/978-981-16-3420-8_5

as mobile phones and sensors, it becomes efficient and beneficial to learn machine learning models over distributed decentralized edge devices, without moving the data to data centers. This scenario of machine learning is called federated learning. Many works have studied different aspects of federated learning, including data privacy [1–5], data heterogeneity [6, 7], communication efficiency [8], and building distributed system.[1]

Automated Machine Learning System AutoML (automatic machine learning) is an attractive research topic that automatically tunes various configurations in machine learning workloads. Previous works have extensively studied automatic approaches for feature selection [9, 10], feature generation [11, 12], model selection [13–15], and hyperparameter tuning [16–19]. There are also some efforts toward the distributed execution of machine learning models, such as synchronization interval and allocated resources [20–22].

References

1. Naehrig, Michael and Lauter, Kristin and Vaikuntanathan, Vinod: Can homomorphic encryption be practical? Proceedings of the 3rd ACM workshop on Cloud computing security workshop. 113–124 (2011)
2. Gentry, Craig and Boneh, Dan: A fully homomorphic encryption scheme. Stanford university. 20(9) (2009)
3. Gentry, Craig: Fully homomorphic encryption using ideal lattices. Proceedings of the Forty-first Annual ACM Symposium on Theory of Computing. 169–178 (2009)
4. Bonawitz, Keith and Ivanov, Vladimir and Kreuter, Ben and Marcedone, Antonio and McMahan, H Brendan and Patel, Sarvar and Ramage, Daniel and Segal, Aaron and Seth, Karn: Practical secure aggregation for privacy-preserving machine learning. Proceedings of the 2017 ACM SIGSAC Conference on Computer and Communications Security. 1175–1191 (2017)
5. Hu, Lingxuan and Evans, David: Secure aggregation for wireless networks. Proceedings of the 2003 Symposium on Applications and the Internet Workshops 2003. 384–391 (2003)
6. McMahan, Brendan and Moore, Eider and Ramage, Daniel and Hampson, Seth and y Arcas, Blaise Aguera: Communication-efficient learning of deep networks from decentralized data. Artificial Intelligence and Statistics. 1273–1282 (2017)
7. Li, Tian and Sahu, Anit Kumar and Zaheer, Manzil and Sanjabi, Maziar and Talwalkar, Ameet and Smith, Virginia: Federated optimization in heterogeneous networks. arXiv preprint arXiv:1812.06127. (2018)
8. Konečnỳ, Jakub and McMahan, H Brendan and Yu, Felix X and Richtárik, Peter and Suresh, Ananda Theertha and Bacon, Dave: Federated learning: Strategies for improving communication efficiency. arXiv preprint arXiv:1610.05492. (2016)
9. Li, Jundong and Cheng, Kewei and Wang, Suhang and Morstatter, Fred and Trevino, Robert P and Tang, Jiliang and Liu, Huan: Feature selection: A data perspective. ACM Computing Surveys (CSUR). 50(6), 1–45 (2017)
10. Guyon, Isabelle and Elisseeff, André: An introduction to variable and feature selection. Journal of Machine Learning Research.3, 1157–1182 (2003)

[1] https://github.com/FederatedAI/FATE.

11. Katz, Gilad and Shin, Eui Chul Richard and Song, Dawn: Explorekit: Automatic feature generation and selection. 2016 IEEE 16th International Conference on Data Mining (ICDM). 979–984 (2016)
12. Kanter, James Max and Veeramachaneni, Kalyan: Deep feature synthesis: Towards automating data science endeavors. 2015 IEEE international conference on data science and advanced analytics (DSAA). 1–10 (2015)
13. van Rijn, Jan N and Abdulrahman, Salisu Mamman and Brazdil, Pavel and Vanschoren, Joaquin: Fast algorithm selection using learning curves. International Symposium on Intelligent Data Analysis. 298–309 (2015)
14. Farahmand, Amir-massoud and Szepesvári, Csaba: Model selection in reinforcement learning. Machine Learning. 85(3), 299–332 (2011)
15. Zucchini, Walter: An introduction to model selection. Journal of Mathematical Psychology. 44(1), 41–61 (2000)
16. Snoek, Jasper and Larochelle, Hugo and Adams, Ryan P: Practical Bayesian optimization of machine learning algorithms. Advances in Neural Information Processing Systems. 25, 2951–2959 (2012)
17. Eggensperger, Katharina and Feurer, Matthias and Hutter, Frank and Bergstra, James and Snoek, Jasper and Hoos, Holger and Leyton-Brown, Kevin: Towards an empirical foundation for assessing Bayesian optimization of hyperparameters. NIPS workshop on Bayesian Optimization in Theory and Practice. 10, 3 (2013)
18. Bergstra, James and Bardenet, Rémi and Bengio, Yoshua and Kégl, Balázs: Algorithms for hyper-parameter optimization. Advances in Neural Information Processing Systems. 24, 2546–2554 (2011)
19. Bergstra, James and Bengio, Yoshua: Random search for hyper-parameter optimization. The Journal of Machine Learning Research. 13(1), 281–305 (2012)
20. Wang, Jianyu and Joshi, Gauri: Adaptive communication strategies to achieve the best error-runtime trade-off in local-update SGD. arXiv preprint arXiv:1810.08313. (2018)
21. Liu, Chris and Zhang, Pengfei and Tang, Bo and Shen, Hang and Zhu, Lei and Lai, Ziliang and Lo, Eric: Towards self-tuning parameter servers. arXiv preprint arXiv:1810.02935. (2018)
22. Sparks, Evan R and Talwalkar, Ameet and Haas, Daniel and Franklin, Michael J and Jordan, Michael I and Kraska, Tim: Automating model search for large scale machine learning. Proceedings of the Sixth ACM Symposium on Cloud Computing. 368–380 (2015)

Printed in the United States
by Baker & Taylor Publisher Services